We live in the information society. The main aim of this book is to describe the basic ideas of information theory, answering questions such as how we may transmit and store information as compactly as possible, what is the maximum quantity of information that can be transmitted by a particular channel or network, and how security can be assured.

This book covers all the basic ideas of information theory and sets them in the context of current applications. These include Shannon's information measure, discrete and continuous information sources and information channels with or wihout memory, source and channel decoding, rate distortion theory, error-correcting codes and the information theoretical approach to cryptology. Throughout the book special attention has been paid to multiterminal or network information theory, an area where there are still many unanswered questions, but which is of great significance because most information is transmitted by networks. Each chapter concludes with questions and worked solutions.

This book will be of interest to advanced undergraduates and graduate students in electrical engineering, computer science, informatics, mathematics, physics and management sciences.

Information Theory

Information Theory

Jan C. A. van der Lubbe

Associate Professor
Information Theory Group
Department of Electrical Engineering
Delft University of Technology

Translated by Hendrik Jan Hoeve and Steve Gee

CAMBRIDGE
UNIVERSITY PRESS

PUBLISHED BY THE PRESS SYNDICATE OF THE UNIVERSITY OF CAMBRIDGE
The Pitt Building, Trumpington Street, Cambridge CB2 1RP, United Kingdom

CAMBRIDGE UNIVERSITY PRESS
The Edinburgh Building, Cambridge CB2 2RU, United Kingdom
40 West 20th Street, New York, NY 10011-4211, USA
10, Stamford Road, Oakleigh, Melbourne 3166, Australia

Originally published in Dutch as *Informatietheorie* by VSSD and © VSSD
1988

First published in English by Cambridge University Press 1997 as
Information Theory

Printed in the United Kingdom at the University Press, Cambridge

Typeset in Times

A catalogue record for this book is available from the British Library

ISBN 0 521 46198 7 hardback
ISBN 0 521 46760 8 paperback

Q360

·L83413

1997

Contents

Preface

On all levels of society systems have been introduced that deal with the transmission, storage and processing of information. We live in what is usually called the infomation society. Information has become a key word in our society. It is not surprising therefore that from all sorts of quarters interest has been shown in what information really is and consequently in acquiring a better knowledge as to how information can be dealt with as efficiently as possible.

Information theory is characterized by a quantitative approach to the notion of information. By means of the introduction of measures for information answers will be sought to such questions as: How to transmit and store information as compactly as possible? What is the maximum quantity of information that can be transmitted through a channel? How can security best be arranged? Etcetera. Crucial questions that enable us to enhance the performance and to grasp the limits of our information systems.

This book has the purpose of introducing a number of basic notions of information theory and clarifying them by showing their significance in present applications. Matters that will be described are, among others: Shannon's information measure, discrete and continuous information sources and information channels with or without memory, source and channel decoding, rate distortion theory, error-correcting codes and the information theoretical approach to cryptology. Special attention has been paid to multiterminal or network information theory; an area with still lots of unanswered questions, but which is of great significance because most of our information is transmitted by networks.

All chapters are concluded with questions and worked solutions. That makes the book suitable for self study.

The content of the book has been largely based on the present lectures by the author for students in Electrical Engineering, Technical Mathematics and Informatics, Applied Physics and Mechanical Engineering at the Delft University of Technology, as well as on former lecture notes by Profs. Ysbrand Boxma, Dick Boekee and Jan Biemond. The questions have been derived from recent exams.

The author wishes to express his gratitude to the colleagues mentioned above as well as the other colleagues who in one way or other contributed to this textbook. Especially I wish to thank E. Prof. Ysbrand Boxma, who lectured on information theory at the Delft University of Technology when I was a student and who introduced me to information theory. Under his inspiring guidance I received my M.Sc. in Electrical Engineering and my Ph.D. in the technical sciences. In writing this book his old lecture notes were still very helpful to me. His influence has been a determining factor in my later career.

Delft, December 1996   Jan C.A. van der Lubbe

1

Discrete information

1.1 The origin of information theory

Information theory is the science which deals with the concept 'information', its measurement and its applications. In its broadest sense distinction can be made between the American and British traditions in information theory.

In general there are three types of information:
- *syntactic information*, related to the symbols from which messages are built up and to their interrelations,
- *semantic information*, related to the meaning of messages, their referential aspect,
- *pragmatic information*, related to the usage and effect of messages.

This being so, syntactic information mainly considers the form of information, whereas semantic and pragmatic information are related to the information content.

Consider the following sentences:
(i)   John was brought to the railway station by taxi.
(ii) The taxi brought John to the railway station.
(iii) There is a traffic jam on highway A3, between Nuremberg and Munich in Germany.
(iv) There is a traffic jam on highway A3 in Germany.

The sentences (i) and (ii) are syntactically different. However, semantically and pragmatically they are identical. They have the same meaning and are both equally informative.

The sentences (iii) and (iv) do not differ only with respect to their syntax, but also with respect to their semantics. Sentence (iii) gives more precise information than sentence (iv).

1

The pragmatic aspect of information mainly depends on the context. The information contained in the sentences (iii) and (iv) for example is relevant for someone in Germany, but not for someone in the USA.

The semantic and pragmatic aspects of information are studied in the British tradition of information theory. This being so, the British tradition is closely related to philosophy, psychology and biology. The British tradition is influenced mainly by scientists like MacKay, Carnap, Bar-Hillel, Ackoff and Hintikka.

The American tradition deals with the syntactic aspects of information. In this approach there is full abstraction from the meaning aspects of information. There, basic questions are the measurement of syntactic information, the fundamental limits on the amount of information which can be transmitted, the fundamental limits on the compression of information which can be achieved and how to build information processing systems approaching these limits. A rather technical approach to information remains.

The American tradition in information theory is sometimes referred to as communication theory, mathematical information theory or in short as information theory. Well-known scientists of the American tradition are Shannon, Renyi, Gallager and Csiszár among others.

However, Claude E. Shannon, who published his article "A mathematical theory of communication" in 1948, is generally considered to be the founder of the American tradition in information theory. There are, nevertheless, a number of forerunners to Shannon who attempted to formalise the efficient use of communication systems.

In 1924 H. Nyquist published an article wherein he raised the matter of how messages (or characters, to use his own words) could be sent over a telegraph channel with maximum possible speed, but without distortion. The term information however was not yet used by him as such.

It was R.V.L. Hartley (1928) who first tried to define *a measure of information*. He went about it in the following manner.

Assume that for every symbol of a message one has a choice of s possibilities. By now considering messages of l symbols, one can distinguish s^l messages. Hartley now defined the amount of information as the logarithm of the number of distinguishable messages. In the case of messages of length l one therefore finds

$$H_H(s^l) = \log\{s^l\} = l \log\{s\}. \tag{1.1}$$

For messages of length 1 one would find

$$H_H(s^1) = \log\{s\}$$

and thus

$$H_H(s^l) = l\, H_H(s^1).$$

This corresponds with the intuitive idea that a message consisting of l symbols, by doing so, contains l times as much information as a message consisting of only one symbol. This also accounts for the appearance of the logarithm in Hartley's definition.

It can readily be shown that the only function that satisfies the equation

$$f\{s^l\} = l f\{s\}$$

is given by

$$f\{s\} = \log\{s\}, \tag{1.2}$$

which yields Hartley's measure for the amount of information. Note that the logarithm also guarantees that the amount of information increases as the number of symbols s increases, which is in agreement with our intuition.

The choice of the base of the logarithm is arbitrary and is more a matter of normalisation. If the natural logarithm is used, the unit of information is called the *nat* (natural unit). Usually 2 is chosen as the base. The amount of information is then expressed in *bits* (derived from binary unit, i.e. two-valued unit). In the case of a choice of two possibilities, the amount of information obtained when one of the two possibilities occurs is then equal to 1 bit. It is easy to see that the relationship between bit and nat is given by

$$1 \text{ nat} = 1.44 \text{ bits.}$$

In Hartley's approach as given above, no allowance is made for the fact that the s symbols may have unequal chances of occurring or that there could be a possible dependence between the l successive symbols.

Shannon's great achievement is that he extended the theories of Nyquist and Hartley, and laid the foundation of present-day information theory by associating information with uncertainty using the concept of chance or probability. With regard to Hartley's measure, Shannon proposed that it could indeed be interpreted as a measure for the amount of information, with the assumption that all symbols have an equal probability of occurring. For the general case, Shannon introduced an information measure based on the concept of probability, which includes Hartley's measure as a special case.

Some attention will first be paid to probability theory, during which some useful notations will be introduced, before introducing Shannon's definition of information.

1.2 The concept of probability

Probability theory is the domain dealing with the concept of probability. The starting point of probability theory is that experiments are carried out which then yield certain outcomes. One can also think in terms of an information source which generates symbols. Every occurrence of a symbol can then be regarded as an event. It is assumed that one is able to specify which possible outcomes or events can occur. The collection of all possible outcomes or events is called the *sample space*. It is now possible to speak of the probability that an experiment has a certain outcome, or of the probability that an information source will generate a certain symbol or message. Each event or outcome has a number between 0 and 1 assigned to it, which indicates how large the probability is that this outcome or event occurs. For simplicity it is assumed that the sample space has a finite number of outcomes.

Consider a so-called *probabilistic experiment* X with possible outcomes/events x_i, with $x_i \in X$ and X the probability space as defined by

$$X = \{x_1,\ldots,x_i,\ldots,x_n\}. \tag{1.3}$$

If we think of throwing a die, then x_1 could be interpreted as the event that "1" is thrown, x_2 the event that "2" is thrown, etc. In the case of the die it is obvious that $n = 6$.

Each event will have a certain probability of occurring. We denote the probability related to x_i by $p(x_i)$ or simply p_i. The collection of probabilities with regard to X is denoted by

$$P = \{p_1,\ldots,p_i,\ldots,p_n\}, \tag{1.4}$$

and is called the *probability distribution*. The probability distribution satisfies two fundamental requirements:

(i) $p_i \geq 0$, for all i.

(ii) $\displaystyle\sum_{i=1}^{n} p_i = 1.$

That is, no probability can take on a negative value and the sum of all the probabilities is equal to 1.

Sometimes we can discern two types of outcomes in one experiment, such that we have a combination of two subexperiments or subevents. When testing IC's for example, one can pay attention to how far certain requirements are met (well, moderately or badly for example), but also to the IC's type number. We are then in actual fact dealing with two sample spaces, say X and Y, where the sample space Y, relating to experiment Y, is in general terms defined by

$$Y = \{y_1, \ldots, y_j, \ldots, y_m\}, \tag{1.5}$$

and where the accompanying probability distribution is given by

$$Q = \{q_1, \ldots, q_j, \ldots, q_m\}, \tag{1.6}$$

where $q(y_j) = q_j$ is the probability of event y_j. We can now regard (X, Y) as a probabilistic experiment with pairs of outcomes (x_i, y_j), with $x_i \in X$ and $y_j \in Y$. The probability $r(x_i, y_j)$, also denoted by r_{ij} or $p(x_i, y_j)$, is the probability that experiment (X, Y) will yield (x_i, y_j) as outcome and is called the *joint probability*. If the joint probability is known, one can derive the probabilities p_i and q_j, which are then called the *marginal probabilities*. It can be verified that for all i

$$p_i = \sum_{j=1}^{m} r_{ij}, \tag{1.7}$$

and for all j

$$q_j = \sum_{i=1}^{n} r_{ij}. \tag{1.8}$$

Since the sum of all the probabilities p_i must be equal to 1 (and likewise the sum of the probabilities q_j), it follows that the sum of the joint probabilities must also be equal to 1:

$$\sum_{i=1}^{n} \sum_{j=1}^{m} r_{ij} = 1.$$

Besides the joint probability and the related marginal probability, there is a third type, namely the *conditional probability*. This type arises when a probabilistic experiment Y is conditional for experiment X. That is, if the probabilities of the outcomes of X are influenced by the outcomes of Y. We are then interested in the probability of an event, x_i for example, given that another event, y_j for example, has already occurred.

Considering the words in a piece of English text, one may ask oneself, for example, what the probability is of the letter "n" appearing if one has already received the sequence "informatio". The appearances of letters in words often depend on the letters that have already appeared. It is very unlikely, for example, that the letter "q" will be followed by the letter "t", but much more likely that the letter "u" follows.

The conditional probability of x_i given y_j is defined as

$$p(x_i/y_j) = \frac{r(x_i,y_j)}{q(y_j)}, \quad \text{provided } q(y_j) > 0,$$

or in shortened notation

$$p_{ij} = \frac{r_{ij}}{q_j}, \qquad \text{provided } q_j > 0. \tag{1.9}$$

The conditional probability of y_j given x_i can be defined in an analogous manner as

$$q(y_j/x_i) = \frac{r(x_i,y_j)}{p(x_i)}, \quad \text{provided } p(x_i) > 0,$$

or simply

$$q_{ji} = \frac{r_{ij}}{p_i}, \qquad \text{provided } p_i > 0. \tag{1.10}$$

From the definitions given it follows that the joint probability can be written as the product of the conditional and marginal probabilities:

$$r(x_i,y_j) = q(y_j)\, p(x_i/y_j) = p(x_i)\, q(y_j/x_i). \tag{1.11}$$

The definition of conditional probability can be simply extended to more than two events. Consider x_i, y_j and z_k for example:

$$p(x_i,y_j,z_k) = r(y_j,z_k)\, p(x_i/y_j,z_k)$$

$$= p(z_k)\, p(y_j/z_k)\, p(x_i/y_j,z_k),$$

hence

$$p(x_i/y_j,z_k) = p(x_i,y_j,z_k) / r(y_j,z_k).$$

Returning to the conditional probability, summation over the index i with y_j given yields

$$\sum_{i=1}^{n} p(x_i/y_j) = 1. \tag{1.12}$$

Whenever an event y_j has occurred, one of the events in X must also occur. Thus, summation will yield 1. Note that the converse is not true. It is generally true that

$$\sum_{j=1}^{m} p(x_i/y_j) \neq 1. \tag{1.13}$$

A handy aid which will be of use in the following is *Bayes' theorem*. It is often the case that the conditional probability $q(y_j/x_i)$ is known, but that we want to determine the conditional probability $p(x_i/y_j)$. One can do this by making use of the following relations:

$$r(x_i, y_j) = p(x_i)\, q(y_j/x_i) = q(y_j)\, p(x_i/y_j).$$

Hence, if $q(y_j) > 0$,

$$p(x_i/y_j) = \frac{p(x_i)\, q(y_j/x_i)}{q(y_j)},$$

or also

$$p(x_i/y_j) = \frac{p(x_i)\, q(y_j/x_i)}{\sum_{i=1}^{n} p(x_i)\, q(y_j/x_i)}. \tag{1.14}$$

We are thus able to calculate $p(x_i/y_j)$ with the help of $q(y_j/x_i)$.

Finally, a comment about the concept of independence. The situation can arise that

$$p(x_i/y_j) = p(x_i).$$

That is, the occurrence of y_j has no influence on the occurence of x_i. But it then also follows that

$$r(x_i, y_j) = p(x_i)\, q(y_j)$$

and

$$q(y_j/x_i) = q(y_j).$$

In this case one says that the events are independent of each other. The reverse is also true, from $r(x_i,y_j) = p(x_i) \, q(y_j)$ it follows that $q(y_j/x_i) = q(y_j)$ and $p(x_i/y_j) = p(x_i)$. Two experiments X and Y are called *statistically independent* if for all i and j

$$r(x_i,y_j) = p(x_i) \, q(y_j). \tag{1.15}$$

An experiment X is called *completely dependent* on another, Y, if for all j, there is a unique i, say k, such that

$$p(x_k/y_j) = 1, \tag{1.16}$$

or

$$p(x_k,y_j) = p(y_j). \tag{1.17}$$

1.3 Shannon's information measure

As we saw in Section 1.1, Hartley's definition of information did not take the various probabilities of occurrence of the symbols or events into account. It was Shannon who first associated information with the concept of probability.

This association is in actual fact not illogical. If we consider a sample space where all events have an equal probability of occurring, there is great uncertainty about which of the events will occur. That is, when one of these events occurs it will provide much more information than in the cases where the sample space is structured in such a way that one event has a large probability of occurring. Information is linked to the concept of chance via uncertainty.

Before considering to what extent *Shannon's information measure* satisfies the properties one would in general expect of an information measure, we first give his definition.

Definition 1.1
Let X be a probabilistic experiment with sample space X and probability distribution P, where $p(x_i)$ or p_i is the probability of outcome $x_i \in X$. Then the average amount of information is given by

$$H(X) = - \sum_{i=1}^{n} p(x_i) \log p(x_i) = - \sum_{i=1}^{n} p_i \log p_i. \tag{1.18}$$

○

Other notations for Shannon's information measure are $H(X)$, $H(P)$ and $H(p_1,\dots,p_n)$. All of these notations will be used interchangeably in this text.

Because this measure for the amount of information is attended with the choice (selection) from n possibilities, one sometimes also speaks of the measure for the amount of *selective information*.

Because 2 is usually chosen as the base, the unit of information thereby becoming the bit, this will not be stated separately in future, but left out.

In the case of two outcomes with probabilities $p_1 = p$ and $p_2 = 1 - p$ we find

$$H(P) = -p \log p - (1 - p) \log (1 - p). \tag{1.19}$$

Figure 1.1 shows how $H(P)$ behaves as a function of p. It can be concluded that if an outcome is certain, that is, occurs with a probability of 1, the information measure gives 0. This is in agreement with the intuitive idea that certain events provide no information. The same is true for $p = 0$; in that case the other outcome has a probability of 1.

When $p = 0.5$, $H(P)$ reaches its maximum value, which is equal to 1 bit. For $p = 0.5$, both outcomes are just as probable, and one is completely uncertain about the outcome. The occurrence of one of the events provides the maximum amount of information in this case.

As an aside, note that by definition $0 \cdot \log(0) = 0$.

Returning to the general case, we can posit that the information measure satisfies four intuitive requirements:

I $H(P)$ is *continuous* in p

Figure 1.1. $H(P) = H(p, 1 - p)$ as a function of p.

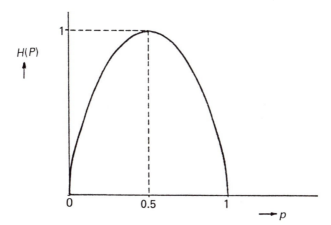

II $H(P)$ is *symmetric*. That is, the ordering of the probabilities $p_1,\ldots,p_n$ does not influence the value of $H(P)$.

III $H(P)$ is *additive*. If X and Y are two sample spaces, where outcomes in X are independent of those in Y, then we find for the information relating to joint events (x_i, y_j)

$$H(p_1 q_1, \ldots, p_1 q_m, \ldots, p_n q_1, \ldots, p_n q_m)$$

$$= H(p_1, \ldots, p_n) + H(q_1, \ldots, q_m). \tag{1.20}$$

IV $H(P)$ is maximum if all probabilities are equal. This corresponds with the situation where maximum uncertainty exists. $H(P)$ is minimum if one outcome has a probability equal to 1.

A short explanation of a number of the above requirements follows.

Ad II. That Shannon's information measure is symmetric means that changing the sequence in which one substitutes the probabilities does not change the amount of information. A consequence of this is that different sample spaces with probability distributions that have been obtained from the permutations of a common probability distribution will result in the same amount of information.

Example 1.1
Consider the experiments X and Y with the following sample spaces:

$$X = \{\text{it will rain tomorrow, it will be dry tomorrow}\}$$
$$\text{where } P = \{0.8, \ 0.2\}$$

and

$$Y = \{\text{John is younger than 30, John is at least 30}\}$$
$$\text{where } Q = \{0.2, \ 0.8\}.$$

The amount of information with relation to X is

$$H(X) = -0.8 \log 0.8 - 0.2 \log 0.2 = 0.72 \text{ bit}$$

and with relation to Y

$$H(Y) = -0.2 \log 0.2 - 0.8 \log 0.8 = 0.72 \text{ bit}$$

and thus

$$H(X) = H(Y). \qquad \triangle$$

From this example, it can be concluded that Shannon's information measure is not concerned with the contents of information. The probabilities with which events occur are of importance and not the events themselves.

Ad III. That Shannon's information measure satisfies the property formulated in equation (1.20), follows directly by writing it out in terms of the probabilities. The property of additivity is best illustrated by the following example. Consider 2 dice. Because the outcomes of the 2 dice are independent of each other, it will not make any difference whether the dice are thrown at the same time or one after the other. The information related to the dice when thrown together will be the same as the successive information that one obtains by throwing one dice and then the other.

If $H(X)$ is the amount of information in relation to throwing one die and $H(Y)$ the amount of information in relation to throwing the other die (note that in this case $H(X) = H(Y)$) while $H(X,Y)$ is the information in relation to two dice thrown at the same time, then it must follow that

$$H(X,Y) = H(X) + H(Y). \tag{1.21}$$

This is exactly what the additivity property asserts.

Ad IV. That the amount of information will be maximum in the case of equal probabilities is obvious, in view of the fact that the uncertainty is the greatest then and the occurrence of one of the events will consequently yield a maximum of information.

In the following theorem, not only the maximum amount of information but also the minimum amount of information will be determined.

Theorem 1.1

Let $X = (x_1,...,x_n)$ be the sample space of experiment X, while $P = (p_1,...,p_n)$ is the corresponding probability distribution. We then find that

(i) $H(P) \leq \log n$, $\tag{1.22}$
    with equality if and only if $p_i = 1/n$ for all $i = 1,...,n$.
(ii) $H(P) \geq 0$, $\tag{1.23}$
    with equality if and only if there is a k such that $p_k = 1$ while for all
    other $i \neq k$ $p_i = 0$.

Proof

(i) During this proof, use will be made of the following inequality (compare Figure 1.2):

$$\ln a \leq a - 1. \tag{1.24}$$

Now consider $H(P) - \log(n)$:

$$H(P) - \log n = -\sum_{i=1}^{n} p_i \log p_i - \log n = -\sum_{i=1}^{n} p_i \{\log p_i + \log n\}$$

$$= \sum_{i=1}^{n} p_i \log\{\frac{1}{p_i\, n}\}.$$

From the inequality $\ln a \leq a - 1$ it follows that

$$\log a = \frac{\ln a}{\ln 2} \leq (a-1)\frac{\ln e}{\ln 2} = (a-1)\log e. \tag{1.25}$$

Using this inequality leads to

$$H(P) - \log n \leq \sum_{i=1}^{n} p_i \left\{\frac{1}{p_i\, n} - 1\right\} \log e = \left\{\sum_{i=1}^{n} \frac{1}{n} - \sum_{i=1}^{n} p_i\right\} \log e$$

$$= \left\{n\, \frac{1}{n} - 1\right\} \log e = 0. \tag{1.26}$$

It has thus been proven that

$$H(P) \leq \log n,$$

with equality if and only if $1/(p_i\, n) = 1$, which corresponds with $a = 1$ in Figure 1.2. This means that $p_i = 1/n$ for all $i = 1,\ldots,n$.

Figure 1.2. Graphical representation of $\ln a \leq a - 1$.

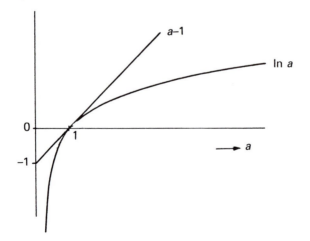

(ii) Because p_i and $-\log p_i$ cannot both be negative, the amount of information is always positive or equal to zero. Hence

$$H(P) \geq 0.$$

It can readily be seen that $H(P)$ can only become equal to 0 if there is one probability in P equal to 1, while all other probabilities are zero.   □

The maximum amount of information is therefore equal to log n. In order to get an impression of the amount of information that an information system delivers, we consider the following example.

Example 1.2
A television image consists of 576 lines, each built up from 720 picture elements. One single television screen image therefore consists of 414 720 picture elements in total. Under the assumption that we are dealing with a grey scale image, where each picture element can display one of 10 intensity intervals, there are 10^{414720} different TV images possible. If each of these images has an equal probability of occurring, the amount of information contained in an image is equal to

$$H(P) = \log n = \log(10^{414720}) \approx 1.4 \cdot 10^6 \text{ bits.} \qquad \triangle$$

Above we have considered a few properties of Shannon's information measure. There are of course still other properties which can be derived regarding this information measure. These will be considered in the coming chapters.

- We have seen that the amount of information does not change if the probabilities are substituted in a different order. Let us now consider the following two probability distributions:

    $$P = \{0.50, \ 0.25, \ 0.25\}$$
    and
    $$Q = \{0.48, \ 0.32, \ 0.20\}.$$

When we calculate the corresponding amount of information for both cases, we find

$$H(P) = H(Q) = 1.5 \text{ bits.}$$

It appears that different probability distributions can lead to the same amount of information: some experiments can have different probability distributions, but an equal amount of information. Figure 1.3 shows geometrically which probability distributions lead to the same amount of information for 3 probabilities ($n = 3$).

The closed curves indicate the probability distributions that lead to the same amount of information. The corresponding values of the probabilities for each arbitrary point on a curve can be found by projection on the lines of p_1, p_2 and p_3.

It is easy to verify that the maximum amount of information for $n = 3$ is equal to

$$H(P) = \log 3 = 1.58 \text{ bits.}$$

Since there is but one probability distribution that can lead to the maximum amount of information, namely $P = \{\frac{1}{3}, \frac{1}{3}, \frac{1}{3}\}$, we find in that case precisely one point in Figure 1.3 instead of a closed curve.

- In order to get more insight into what Shannon's information measure represents, we consider the following two examples.

Example 1.3

Suppose we have a graphical field consisting of 16 regions, one of which is shaded (see Figure 1.4).

By asking questions which can only be answered with a yes or a no, we have to determine where the shaded region is situated. What is the best strategy? One can guess, but then one takes the risk of having to ask 16 questions before finally finding the shaded region. It is better to work selectively. The question and answer game could then end up looking like the following, for

Figure 1.3. Curves with an equal amount of information in the case of a sample space of 3.

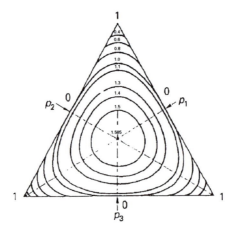

example (see also Figure 1.5):

1. Is the shaded region one of the bottom eight regions?
   Answer: "no", thus regions 9 to 16 can be dismissed.
2. Is the shaded region one of the four regions remaining to the left?
   Answer: "yes", thus the shaded region is 1, 2, 5 or 6.
3. Is the shaded region one of the bottom two of the four remaining regions?
   Answer: "yes", thus the shaded region is 5 or 6.
4. Is it the left region?
   Answer: "no", so the shaded region is 6.

There are therefore four questions necessary in total to determine which of the 16 regions is shaded.

If we now consider the amount of information regarding this problem we find, as all 16 regions are equally probable, that

$$H(P) = -\sum_{i=1}^{16} \frac{1}{16} \log \frac{1}{16} = \log(16) = 4 \text{ bits.}$$

The amount of information apparently corresponds with the minimum number of questions that one must ask to determine which outcome (the shaded region in this case) has occurred. △

In the following example, we examine if the interpretation of Example 1.3 is also valid when dealing with probability distributions where not all probabilities are equal.

Figure 1.4. Question and answer game: finding the shaded region.

Figure 1.5. Example of a tree structure for a question and answer game.

1	2	3	4
5	6	7	8
9	10	11	12
13	14	15	16

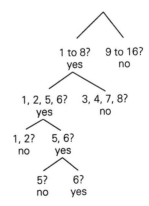

Example 1.4

A sample space X is given by $X = \{x_1, x_2, x_3\}$, while the accompanying probability space is given by $P = \{1/2, 1/4, 1/4\}$. Playing the "yes" and "no" game again, it seems obvious to ask for x_1 first, as this outcome has the greatest probability.

If the answer is "yes", then we have found the outcome in one go. If the answer is "no", then the outcome is obviously x_2 or x_3. To determine if it is x_2 or x_3 costs another question, so that we need to ask two questions in total to know the outcome.

One must therefore ask either one question or two questions, with equal probabilities, hence the average is 1.5 questions.

If we calculate the amount of information according to Shannon, then we find:

$$H(P) = -\frac{1}{2}\log\frac{1}{2} - \frac{1}{4}\log\frac{1}{4} - \frac{1}{4}\log\frac{1}{4} = 1.5 \text{ bits.}$$

The previously given interpretation is therefore also valid for unequal probabilities.   △

1.4 Conditional , joint and mutual information measures

In Section 1.2 we referred to a probabilistic experiment (X,Y) with possible outcomes (x_i, y_j), where $(x_i, y_j) \in (X,Y)$.

On the basis of the size of the sample space (X,Y) it can be concluded that experiment (X,Y) has nm possible joint outcomes in total. If we now want to define the amount of information with regard to (X,Y), then the following course can be followed.

There are nm joint events (x_i, y_j), with probabilities of occurring $r(x_i, y_j)$ or r_{ij} (see Section 1.2 for notation). Now suppose that we write the nm joint events as events $z_1, z_2, \ldots, z_{nm}$ and the corresponding probabilities as $p(z_1), p(z_2), \ldots, p(z_{nm})$. We then in actual fact have a one-dimensional sample space again, and with the definition of the marginal information measure we find

$$H(Z) = -\sum_{k=1}^{nm} p(z_k) \log p(z_k). \tag{1.27}$$

But because each $p(z_k)$ will be equal to one of the probabilities $r(x_i, y_j)$, summation over k will yield the same as summation over i and j. In other words:

$$H(Z) = -\sum_{i=1}^{n}\sum_{j=1}^{m} r(x_i,y_j) \log\left[r(x_i,y_j)\right].$$

This leads to the following definition of the joint information measure.

Definition 1.2
Consider a probabilistic experiment (X,Y) with two-dimensional sample space (X,Y), where r_{ij} or $r(x_i,y_j)$ is the probability of x_i and y_j, then the *joint information measure* is defined as

$$H(X,Y) = -\sum_{i=1}^{n}\sum_{j=1}^{m} r(x_i,y_j) \log\left[r(x_i,y_j)\right]. \tag{1.28}$$

○

We will use the alternative notations $H(R)$ and $H(r_{11},...,r_{nm})$, as well as $H(X,Y)$, interchangeably.

So far we have seen that the marginal information measure can be defined on the basis of marginal probabilities and that joint probabilities lead to the introduction of the joint information measure. We will now consider if a conditional information measure can be defined in relation to conditional probabilities.

The probabilistic experiments X and Y are being considered again. Now suppose that we are interested in the amount of information with regard to Y under the condition that outcome x_i has already occurred. We then have probabilities $q(y_j/x_i)$, $j = 1,...,m$, instead of probabilities $q(y_j)$, $j = 1,...,m$, but with the sum still equal to 1.

The amount of information with regard to Y given outcome x_i can then, in analogy with the marginal information measure, be defined as

$$H(Y/x_i) = -\sum_{j=1}^{m} q(y_j/x_i) \log\left[q(y_j/x_i)\right]. \tag{1.29}$$

By now averaging over all values x_i, the average amount of information of Y given foreknowledge of X is found:

$$\sum_{i=1}^{n} p(x_i)H(Y/x_i) = \sum_{i=1}^{n} p(x_i)\left\{-\sum_{j=1}^{m} q(y_j/x_i) \log[q(y_j/x_i)]\right\}$$

$$= -\sum_{i=1}^{n}\sum_{j=1}^{m} p(x_i)\, q(y_j/x_i) \log[q(y_j/x_i)]$$

$$= -\sum_{i=1}^{n} \sum_{j=1}^{m} r(x_i,y_j) \log[q(y_j/x_i)].$$

This quantity is denoted by the conditional amount of information $H(Y/X)$. This leads to the following definition.

Definition 1.3
The *conditional information measure* with regard to experiment Y given X is equal to

$$H(Y/X) = -\sum_{i=1}^{n} \sum_{j=1}^{m} r(x_i,y_j) \log[q(y_j/x_i)]. \tag{1.30}$$

In an analogous manner, one can define the amount of information that is obtained on average with regard to X if Y is known as

$$H(X/Y) = -\sum_{i=1}^{n} \sum_{j=1}^{m} r(x_i,y_j) \log[p(x_i/y_j)]. \tag{1.31}$$

○

Instead of $H(Y/X)$ and $H(X/Y)$ we will also use the notations related to their spaces: $H(Y/X)$ and $H(X/Y)$

The following theorem gives the minimum and maximum values for $H(Y/X)$.

Theorem 1.2
Let $H(Y/X)$ be the conditional information measure for Y given X, then

(i) $H(Y/X) \geq 0,$ (1.32)

(ii) $H(Y/X) \leq H(Y),$ (1.33)
            with equality if X and Y are stochastically independent.

Proof
(i) Because $q(y_j/x_i) \leq 1$ for all i and j, it follows that $\{-\log p(y_j/x_i)\} \geq 0$, so that it follows directly from the definition that

    $$H(Y/X) \geq 0.$$

(ii) $$H(Y/X) - H(Y) = -\sum_{i=1}^{n} \sum_{j=1}^{m} r(x_i,y_j) \log[q(y_j/x_i)] + \sum_{j=1}^{m} q(y_j) \log q(y_j)$$

$$= \sum_{i=1}^{n} \sum_{j=1}^{m} r(x_i, y_j) \log \left[\frac{q(y_j)}{q(y_j/x_i)} \right].$$

Using the previously mentioned inequality $\ln(a) \le a - 1$ (see Figure 1.2) it follows that

$$H(Y/X) - H(Y) \le \sum_{i=1}^{n} \sum_{j=1}^{m} r(x_i, y_j) \left[\frac{q(y_j)}{q(y_j/x_i)} - 1 \right] \log e.$$

The right-hand side of this inequality can be written as

$$\sum_{i=1}^{n} \sum_{j=1}^{m} p(x_i) q(y_j/x_i) \frac{q(y_j)}{q(y_j/x_i)} \log e - \sum_{i=1}^{n} \sum_{j=1}^{m} r(x_i, y_j) \log e$$

$$= \log e - \log e = 0.$$

Hence

$$H(Y/X) \le H(Y).$$

The two amounts of information are equal if $q(y_j) = q(y_j/x_i)$ for all i and j, as is the case with stochastic independence. $\square$

The conclusion that can be attached to this theorem is that (on average) the conditional amount of information is always less than or equal to the marginal amount of information. In other words, information about X will generally lead to a reduction of uncertainty. This is in agreement with our intuitive ideas about foreknowledge.

A direct relationship exists between the marginal, conditional and joint information measures, as shown by the following theorem.

Theorem 1.3
For all experiments X and Y,

$$H(X,Y) = H(X) + H(Y/X)$$

$$= H(Y) + H(X/Y). \tag{1.34}$$

Proof
We have

$$H(X,Y) = -\sum_{i=1}^{n} \sum_{j=1}^{m} r(x_i, y_j) \log[p(x_i)\, q(y_j/x_i)]$$

$$= -\sum_{i=1}^{n}\sum_{j=1}^{m} r(x_i,y_j) \log p(x_i) -\sum_{i=1}^{n}\sum_{j=1}^{m} r(x_i,y_j) \log[q(y_j/x_i)]$$

$$= H(X) + H(Y/X).$$

The proof of $H(X,Y) = H(Y) + H(X/Y)$ proceeds in an analogous manner.   □

What the theorem is really saying is that the joint amount of information is the sum of the amount of information with regard to X and the conditional amount of information of Y given X.

On the basis of Theorem 1.2 and Theorem 1.3, it can furthermore be derived that

$$H(X,Y) = H(X) + H(Y/X) \leq H(X) + H(Y), \tag{1.35}$$

with equality if X and Y are independent.

One can thus suppose that the joint amount of information is maximum if the two probabilistic experiments are independent and decreases as the dependence increases. With absolute dependence, the outcome of Y is known if the outcome of X is known, so that $H(Y/X) = 0$. In that case $H(X,Y) = H(X)$.

One last definition now remains to be given in this section, that of the mutual information measure, which will play an important role with respect to the concept of the capacity of a communication channel as discussed later.

Definition 1.4
The *mutual information measure* with regard to X and Y is defined by

$$I(X;Y) = H(Y) - H(Y/X)$$

$$= \sum_{i=1}^{n}\sum_{j=1}^{m} r(x_i,y_j) \log\left[\frac{r(x_i,y_j)}{p(x_i)\, q(y_j)}\right]. \tag{1.36}$$

○

$I(X;Y)$ can be interpreted as a measure for the dependence between Y and X. When X and Y are independent, $I(X;Y)$ is minimum, namely

$$I(X;Y) = 0.$$

If Y is completely dependent on X, then $H(Y/X) = 0$ and $I(X;Y)$ attains its maximum value which is equal to

$$I(X;Y) = H(Y).$$

It is left to the reader to show that for all X and Y

$$I(X;Y) = H(X) - H(X/Y)$$

$$= H(X) + H(Y) - H(X,Y), \qquad (1.37)$$

and that $I(X;Y)$ is symmetric, that is for all X and Y

$$I(X;Y) = I(Y;X). \qquad (1.38)$$

In this section we have introduced three information measures, namely the conditional, joint and mutual information measures. Some attention was also paid to the different relationships between the measures themselves. These are best illustrated and summarised by the Venn diagrams shown in Figure 1.6.
We have

$$I(X;Y) = H(X) \cap H(Y),$$

$$H(X,Y) = H(X) \cup H(Y).$$

From Figure 1.6(a), the most general case, it can be concluded that:

- $H(X/Y) \leq H(X)$ and $H(Y/X) \leq H(Y)$;

- $I(X;Y) \leq H(Y)$ and $I(X;Y) \leq H(X)$;

Figure 1.6. Relationships between information measures: (a) the general case; (b) the case of independence.

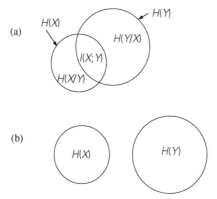

- $I(X;Y) = H(X) - H(X/Y) = H(Y) - H(Y/X);$

- $H(X,Y) = H(X/Y) + I(X;Y) + H(Y/X)$

$$= H(Y) + H(X/Y) = H(X) + H(Y/X);$$

- $H(X,Y) \leq H(X) + H(Y).$

In Figure 1.6(b), X and Y are independent. Note that now:

- $I(X;Y) = 0;$

- $H(X,Y) = H(X) + H(Y);$

- $H(X) = H(X/Y)$ and $H(Y) = H(Y/X).$

The relationships between the various information measures derived in this and previous sections can readily be demonstrated through the use of Venn-diagrams.

1.5 Axiomatic foundations

In Section 1.3, Shannon's information measure was introduced and some properties of the information measure were derived. These properties appeared to correspond with the properties that one would intuitively expect of an information measure. In the uniform case Shannon's information measure equals the information measure of Hartley, which is the logarithm of the number of messages (compare equation (1.1)). Shannon's information measure based on probabilities can be derived directly from the uniform case and the measure of Hartley.

Assume that an information measure should satisfy the following three requirements.

(i)   If all outcomes are split up into groups, then all the values of *H* for the various groups, multiplied by their statistical weights, should lead to the overall *H*.

(ii) *H* should be continuous in p_i.

(iii) If all p_i's are equal, i.e. for all i $p_i = \frac{1}{n}$, then *H* will increase monotonically as a function of *n*. That means the uncertainty will increase for an increasing number of equal probabilities.

For *n* equally probable outcomes *H* should satisfy $H = \log n$ according to Hartley and requirement (iii). For unequal probabilities consider the following case. Assume probabilities $\frac{3}{6}$, $\frac{2}{6}$ and $\frac{1}{6}$. Figure 1.7(a) gives the decision tree for which *H* should be computed.

The value of H with respect to the decision tree of Figure 1.7(c) should be equal to $H^c = \log 6$ (compare requirement (iii)). However, since both decision trees of Figure 1.7(b) and 1.7(c) are fundamentally identical, the value with respect to the decision tree of Figure 1.7(b) should also be equal to $\log 6$; $H^b = \log 6$. On the basis of requirement (i) this value should equal the uncertainty concerning the choice between the branches denoted by $\frac{3}{6}$, $\frac{2}{6}$ and $\frac{1}{6}$ in Figure 1.7(b) (i.e. the H^a searched for) plus the uncertainties with respect to the subbranches multiplied by their weights.
It follows that

$$H^a + \frac{3}{6}\log 3 + \frac{2}{6}\log 2 + \frac{1}{6}\log 1 = \log 6$$

and

$$H^a = -\frac{1}{2}\log\frac{1}{2} - \frac{1}{3}\log\frac{1}{3} - \frac{1}{6}\log\frac{1}{6}.$$

More generally it follows that

$$H(X) = -\sum_{i=1}^{n} p_i \log p_i.$$

Chaundy and McLeod (1960) have given the following characterisation theorem which uniquely determines Shannon's information measure.

Theorem 1.4
Consider a function $f(X) = f(P) = f(p_1,...,p_n)$ and a function $g(.)$, which satisfy the following properties:

(i) $\qquad f(P) = \sum_{i=1}^{n} g(p_i),$

(ii) $\qquad f(.)$ is continuous in the interval [0,1],

(iii) $\qquad f(P)$ is additive,

Figure 1.7. Decision trees related to unequally probable outcomes.

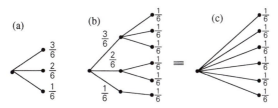

$$f(p_1 q_1, \ldots, p_n q_m) = f(p_1, \ldots, p_n) + f(q_1, \ldots, q_m),$$

(iv) $f(\frac{1}{2}, \frac{1}{2}) = 1;$

then

$$f(P) = H(P) = - \sum_{i=1}^{n} p_i \log p_i. \qquad \square$$

It can be derived from the theorem that it is actually the additivity property that uniquely determines Shannon's information measure.

As an aside, note that since Shannon introduced his information measure in 1948, a variety of research has been carried out searching for alternatives to Shannon's information measure. Particular mention must be made here of the works of Renyi (1960), Daroczy (1970) and Azimoto (1971). With regard to the latter two, the strong requirement of additivity is replaced by a weaker form of additivity. In Van der Lubbe (1981) all of these measures have been brought together in one unifying framework.

1.6 The communication model

The information of a source will generally not in itself be further used. For historical reasons, it is common practice to speak of the manner in which information is used in terms of a *communication model*. In the case of the communication model, the emphasis lies on the transport of information, as generated by a source, to a destination. The storage of information in a memory is likewise of great importance nowadays and although this is not a transmission problem, it can be described in those terms.

During the transport of information, communication takes place between the source which generates information, often called the transmitter or sender, on one side and the destination or receiver on the other side. The basic model is depicted in Figure 1.8. A fundamental problem with the communication between sender and receiver is that errors or distortions can arise during transport through the communication channel as a result of e.g. noise which acts upon the channel. The transport of information must be error-free to a certain degree, depending on the requirements imposed by the receiver. It must therefore be possible to correct errors, or the transport must be good enough that certain errors which are considered to be less serious can be tolerated.

A perfect, i.e. error-free, transmission is not really possible when transmitting such signals as speech, music or video, and one will only be able to impose requirements on the extent to which the received signal differs from the transmitted signal. The required quality leads to the choice of a suitable transmission medium, but especially imposes boundary conditions on the adaptation of this channel to the sender and receiver.

A more detailed description of the communication model is given in Figure 1.9. The communication system should transmit the information generated by the source to the destination as accurately as possible. It is assumed that the information source and the destination as well as the channel are all given. The noise source acting upon the channel is also regarded as given. We assume that the continuous channel transmits signals the nature of which depends on the available physical transmission or storage medium (electric, magnetic, optic) and on the chosen method of modulation. The physical characteristics of a continuous channel such as bandwidth and signal-to-noise ratio are also regarded as given. The purpose of the sender is to make the information from the information source suitable for transportation through the communication channel, while the receiver attempts to correct

Figure 1.8. Elementary communication model.

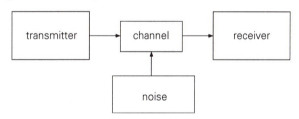

Figure 1.9. Detailed communication model.

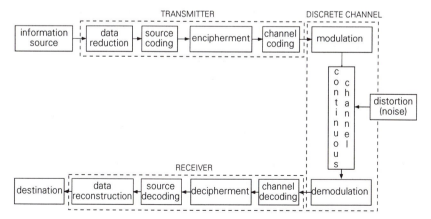

distortions and errors arising in the channel and subsequently transforms the information into a form suitable for the destination.

A subdivision of the sender's functions leads to four sub-aspects. First of all, it will become apparent that not all of the information generated by the information source is relevant for the destination. Due to efficiency considerations, this had best be removed. This is called *data reduction*. The remaining information is called the *effective information*.

This effective information will often have to be processed in another (numerical) way, in binary form for example, and will still contain much internal structure. Through the application of *source coding*, sometimes also called *data compression*, the effective information is represented as compactly as possible.

More and more often it proves to be desirable to secure the resulting information to prevent possible improper use. A solution is to *encipher* the information with the help of secret codes.

The protection of the information against possible errors which can arise in the channel comes forth as the fourth element of the sender. To achieve this, extra information is added which can later be used to reconstruct the original information if any errors have occurred. We speak of *channel coding* when we use codes that detect and/or correct errors.

The information so delivered by the sender is subsequently offered to the channel. We speak of a *discrete channel* if we abstract the channel to a level where it is offered information well separated at the input side and after transmission also produces symbols again at the output side. Internally, this transmission should take place via signals that must be transmitted through a physical medium (the continuous channel). The conversion of the offered symbols into suitable signals takes place via *modulation*. These signals are distorted, in the communication model under consideration, with *noise*. The thus arisen mixture of signal and noise is subsequently converted into symbols again by *demodulation*. Any coincidentally present noise however can have as a result that a presented symbol results in another, incorrect symbol after transmission. The information originating from the channel is now checked for errors and possibly corrected through the use of *channel decoding*.

The resulting information is successively *decrypted* and decoded in the *decoder*. The information is finally brought to the desired form for the destination by means of *data reconstruction*.

The processing steps mentioned here can largely be viewed as deterministic transformations in the sense that the forward and backward transformations yield exactly the original result again. The exceptions to this are data reduction and data reconstruction and the discrete channel wherein presumed stochastic noise appears.

In the following chapters, these aspects of the communication channel will be further worked out and the boundaries wherein one should work in order to achieve efficient and error-free transmission or storage of information will be indicated.

1.7 Exercises

1.1. The sum of the faces of two normal dice when thrown is 7. How much information does this fact supply us with? Explain your answer.
Note: Outcomes such as (6,1) and (1,6) are to be considered as being different.

1.2. A vase contains m black balls and n minus m white balls. Experiment X involves the random drawing of a ball, without it being replaced in the vase. Experiment Y involves the random drawing of a second ball.
a. Determine the amount of information received from experiment X.
b. Determine the amount of information with respect to experiment Y if the colour of the ball selected in experiment X is not known.
c. As question b, now with the assumption that the colour of the ball selected in experiment X is known.

1.3. A roulette wheel is subdivided into 38 numbered compartments of various colours. The distribution of the compartments according to colour is:

2 green,

18 red,

18 black.

The experiment consists of throwing a small ball onto the rotating roulette wheel. The event, that the ball comes to rest in one of the 38 compartments, is equally probable for each compartment.
a. How much information does one receive if one is only interested in the colour?
b. How much information does one receive if one is interested in the colour and the number?
c. What then follows for the amount of conditional information if the colour is known?

1.4. An urn contains 5 black and 10 white balls. Experiment X involves a random drawing of a ball. Experiment Y involves a random drawing of a ball with the ball drawn in experiment X not replaced in the urn. One is interested in the colour of the drawn ball.

a. How much uncertainty does experiment X contain?
b. How large is the uncertainty in experiment Y given that the first ball is black?
c. How large is the uncertainty in experiment Y given that the first ball is white?
d. How much uncertainty does experiment Y contain?

1.5. For a certain exam, 75% of the participating students pass, 25% do not pass. Of the students who have passed, 10% own a car, of those who have failed, 50% own a car.

a. How much information does one receive if one is told the result of a student's exam?
b. How much information is contained in the announcement of a student who has passed that he does or does not have a car?
c. How much uncertainty remains concerning the car ownership of a student if he announces the result of his exam?

1.6. In a certain region 25% of the girls are blond and 75% of all blond girls have blue eyes. Also 50% of all girls have blue eyes. How much information do we receive in each of the following cases:

a. if we know that a girl is blond and we are told the colour (blue/not blue) of her eyes;
b. if we know that a girl has blue eyes and we are told the colour (blond/not blond) of her hair;
c. if we are told both the colour of her hair and that of her eyes.

1.7. Of a group of students, 25% are not suitable for university. As the result of a selection, however, only 75% of these unsuitable students are rejected. 50% of all students are rejected.

a. How much information does one receive if a student, who knows that he is not suitable for university, hears the result of the selection.
b. Answer the same question if the selection is determined by tossing a coin.
c. Compare the results of b with a and give an explanation for the differences.

1.8. Two experiments, X and Y, are given. The sample space with regard to X consists of x_1, x_2 and x_3, that of Y consists of y_1, y_2 and y_3. The joint probabilities $r(x_i, y_j) = r_{ij}$ are given in the following matrix R:

$$\begin{bmatrix} r_{11} & r_{12} & r_{13} \\ r_{21} & r_{22} & r_{23} \\ r_{31} & r_{32} & r_{33} \end{bmatrix} = \begin{bmatrix} \dfrac{7}{24} & \dfrac{1}{24} & 0 \\ \dfrac{1}{24} & \dfrac{1}{4} & \dfrac{1}{24} \\ 0 & \dfrac{1}{24} & \dfrac{7}{24} \end{bmatrix}.$$

a. How much information do you receive if someone tells you the outcome resulting from X and Y?

b. How much information do you receive if someone tells you the outcome of Y?

c. How much information do you receive if someone tells you the outcome of X, while you already knew the outcome of Y?

1.9. A binary communication system makes use of the symbols "zero" and "one". As a result of distortion, errors are sometimes made during transmission. Consider the following events:

u_0 : a "zero" is transmitted;
u_1 : a "one" is transmitted;
v_0 : a "zero" is received;
v_1 : a "one" is received.

The following probabilities are given:

$$p(u_0) = \frac{1}{2}, \ p(v_0/u_0) = \frac{3}{4}, \ p(v_0/u_1) = \frac{1}{2}.$$

a. How much information do you receive when you learn which symbol has been received, while you know that a "zero" has been transmitted?

b. How much information do you receive when you learn which symbol has been received, while you know which symbol has been transmitted?

c. Determine the amount of information that you receive when someone tells you which symbol has been transmitted and which symbol has been received.

d. Determine the amount of information that you receive when someone tells you which symbol has been transmitted, while you know which symbol has been received.

1.8 Solutions

1.1. There are $6^2 = 36$ possible outcomes when throwing two dice, each with the same probability of occurring, namely 1/36. Each throw can therefore deliver an amount of information equal to

$$H(X) = \log 36 = 2.48 \text{ bits/throw.}$$

Since it is given that the sum of the faces is 7, of the 36 possibilities, 6 remain which can still occur, namely 1-6, 6-1, 2-5, 5-2, 3-4, 4-3. Thus, there still exists a remaining uncertainty which can deliver an amount of information equal to

$$H'(X) = \log 6 = 1.24 \text{ bits/throw.}$$

Because it is given that the sum of the faces is 7, the amount of information that this fact gives us is equal to

$$H(X) - H'(X) = \log 36 - \log 6 = \log 6 = 1.24 \text{ bits/throw.}$$

1.2. *a.* The probabilities of drawing a white or a black ball are given by

$$p(w) = \frac{n-m}{n}, \quad p(bl) = \frac{m}{n}.$$

The amount of information that is received from experiment X is therefore:

$$H(X) = -p(w_X) \log p(w_X) - p(bl_X) \log p(bl_X)$$

$$= -\frac{n-m}{n} \log\left[\frac{n-m}{n}\right] - \frac{m}{n} \log \frac{m}{n},$$

where $p(w_X)$ and $p(bl_X)$ are the probabilities that a white and a black ball are drawn respectively.

b. A ball is drawn randomly without replacement, there are thus still $n-1$ balls left over in the vase in the case of experiment Y. A distinction must now be made between the possibilities that a black or a white ball is drawn with experiment X. That is, the conditional probabilities must be determined. These are

Figure 1.10. Conditional probabilities with respect to experiments X and Y of Exercise 1.2.

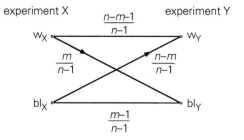

$$p(w_Y/w_X) = \frac{n - m - 1}{n - 1}, \qquad p(bl_Y/bl_X) = \frac{m - 1}{n - 1},$$

$$p(w_Y/bl_X) = \frac{n - m}{n - 1}, \qquad p(bl_Y/w_X) = \frac{m}{n - 1}.$$

See Figure 1.10.

The colour of the ball involved in experiment X is not known, so that there are two possibilities w_X and bl_X. This leads to

$$p(w_Y) = p(w_X)\, p(w_Y/w_X) + p(bl_X)\, p(w_Y/bl_X)$$

$$= \frac{n - m}{n} \frac{n - m - 1}{n - 1} + \frac{m}{n} \frac{n - m}{n - 1} = \frac{n - m}{n} = p(w_X).$$

Similarly, it follows that $p(bl_Y) = p(bl_X)$. It therefore follows for the amount of information resulting from this experiment that $H(Y) = H(X)$. This can also be seen by bearing in mind that experiment X gives no factual information that can decrease the uncertainty over experiment Y.

c. We can distinguish two cases. If a white ball is drawn in experiment X, then the amount of information in experiment Y is

$$H(Y/w_X) = -p(w_Y/w_X) \log p(w_Y/w_X) - p(bl_Y/w_X) \log p(bl_Y/w_X)$$

$$= -\frac{n-m-1}{n-1} \log\left[\frac{n-m-1}{n-1}\right] - \frac{m}{n-1} \log\left[\frac{m}{n-1}\right].$$

If a black ball is drawn, then

$$H(Y/bl_X) = -p(w_Y/bl_X) \log p(w_Y/bl_X) - p(bl_Y/bl_X) \log p(bl_Y/bl_X) =$$

$$= -\frac{n - m}{n - 1} \log\left[\frac{n - m}{n - 1}\right] - \frac{m - 1}{n - 1} \log\left[\frac{m - 1}{n - 1}\right].$$

1.3. *a.* If we observe the colour of the compartment where the ball comes to rest, the experiment can have three possible outcomes, namely green, red and black, with probabilities of occurring $p(\text{green}) = 2/38 = 1/19$, $p(\text{red}) = 18/38 = 9/19$, $p(\text{black}) = 18/38 = 9/19$.

If we are only interested in the colour, we come to an amount of information

$$H(\text{colour}) = -\sum_i p_i \log p_i$$

$$= -\frac{1}{19} \log \frac{1}{19} - 2 \frac{9}{19} \log \frac{9}{19} = -\frac{36}{19} \log 3 + \log 19 = 1.24 \text{ bits.}$$

b. We can determine the amount of information by bearing in mind that the compartment is completely determined by giving the number. There are 38 compartments, each occurring with equal probability, thus

$$H(\text{colour,number}) = H(\text{number}) = \log 38 = 5.25 \text{ bits.}$$

The amount of conditional information $H(\text{colour/number})$ is obviously zero, which is clearly because the colour is automatically known for a given number.

c. The conditional information, if the colour is known, is

$$H(\text{number/colour}) = H(\text{colour,number}) - H(\text{colour})$$

$$= 5.25 - 1.24 = 4.01 \text{ bits.}$$

1.4. *a.* The probabilities of drawing a black or a white ball are $p(\text{bl}_X) = 1/3$, $p(w_X) = 2/3$.
When randomly drawing a ball, we receive an amount of information or uncertainty

$$H(X) = -\frac{1}{3} \log \frac{1}{3} - \frac{2}{3} \log \frac{2}{3} = 0.92 \text{ bit.}$$

b. For experiment Y, the probabilities if the outcome of X is black are $p(\text{bl}_Y/\text{bl}_X) = 4/14 = 2/7$ and $p(w_Y/\text{bl}_X) = 10/14 = 5/7$, so that

$$H(Y/\text{bl}_X) = -\frac{2}{7} \log \frac{2}{7} - \frac{5}{7} \log \frac{5}{7} = 0.86 \text{ bit.}$$

c. If the outcome of X is white, then it similarly follows that $p(\text{bl}_Y/w_X) = 5/14$, $p(w_Y/w_X) = 9/14$, and

$$H(Y/w_X) = -\frac{9}{14} \log \frac{9}{14} - \frac{5}{14} \log \frac{5}{14} = 0.94 \text{ bit.}$$

d. The uncertainty in experiment Y is the weighted sum of the results b and c, namely

$$H(Y/X) = p(\text{bl}_X) H(Y/\text{bl}_X) + p(w_X) H(Y/w_X)$$

$$= \frac{1}{3} \, 0.86 + \frac{2}{3} \, 0.94 = 0.91 \text{ bit.}$$

1.5. *a.* The four possible situations, passing, not passing, owning a car, not owning a car, are denoted by s, $\bar{s}$, c and $\bar{c}$, respectively. When the result of an exam is announced, this delivers an amount of information

$$H(\text{result}) = -p(s) \log p(s) - p(\overline{s}) \log p(\overline{s})$$

$$= -\frac{3}{4} \log \frac{3}{4} - \frac{1}{4} \log \frac{1}{4} = 0.81 \text{ bit.}$$

b. If a student, who has passed, announces whether or not he has a car, then there are two possibilities c and $\overline{c}$ with the given probabilities. Thus

$$H(\text{car owner/passed}) = -p(c/s) \log p(c/s) - p(\overline{c}/s) \log p(\overline{c}/s)$$

$$= -\frac{1}{10} \log \frac{1}{10} - \frac{9}{10} \log \frac{9}{10} = 0.47 \text{ bit.}$$

c. There are four possibilities in total. The corresponding probabilities are

$$p(s,c) = \frac{3}{4} \frac{1}{10} = \frac{3}{40},$$

$$p(s,\overline{c}) = \frac{3}{4} \frac{9}{10} = \frac{27}{40},$$

$$p(\overline{s},c) = \frac{1}{4} \frac{1}{2} = \frac{1}{8},$$

$$p(\overline{s},\overline{c}) = \frac{1}{4} \frac{1}{2} = \frac{1}{8}.$$

The amount of information that is delivered by announcing the result of the exam as well as possible car ownership is then

$$H(\text{car ownership,result}) = -\frac{3}{40} \log \frac{3}{40} - \frac{27}{40} \log \frac{27}{40} - 2 \times \frac{1}{8} \log \frac{1}{8}$$

$$= 1.41 \text{ bits.}$$

The remaining uncertainty about car ownership, if the result of the exam is given, is then

$$H(\text{car ownership/result}) = H(\text{car ownership,result}) - H(\text{result})$$

$$= 1.41 - 0.81 = 0.60 \text{ bit.}$$

One can also obtain this result by calculating the conditional amount of information directly as per the definition for the conditional amount of information.

$$H(\text{car ownership/result}) = \tfrac{3}{4}\left(-\tfrac{1}{10}\log\tfrac{1}{10} - \tfrac{9}{10}\log\tfrac{9}{10}\right)$$
$$+ \tfrac{1}{4}\left(-\tfrac{1}{2}\log\tfrac{1}{2} - \tfrac{1}{2}\log\tfrac{1}{2}\right) = 0.60 \text{ bit.}$$

1.6. *a.* If she is blond, there are two possible eye colours, namely blue and not blue with probabilities $\tfrac{3}{4}$ and $\tfrac{1}{4}$ respectively. One therefore receives a conditional amount of information, that is under the condition that she is blond,

$$H(\text{colour eyes/blond}) = -\tfrac{3}{4}\log\tfrac{3}{4} - \tfrac{1}{4}\log\tfrac{1}{4} = 0.81 \text{ bit.}$$

b. To be able to answer this question we must first determine the probabilities $p(\text{blond/blue})$ and $p(\text{not blond/blue})$. This can be done with the help of Bayes's formula, giving

$$p(\text{blond/blue}) = \frac{p(\text{blond})\, p(\text{blue/blond})}{p(\text{blue})} = \frac{\tfrac{1}{4}\,\tfrac{3}{4}}{\tfrac{1}{2}} = \tfrac{3}{8}.$$

Because a girl is either blond, or not blond, we have

$$p(\text{blond/blue}) + p(\text{not blond/blue}) = 1$$

which gives

$$p(\text{not blond/blue}) = \tfrac{5}{8}.$$

The conditional amount of information that one receives is then

$$H(\text{colour hair/blue}) = -\tfrac{3}{8}\log\tfrac{3}{8} - \tfrac{5}{8}\log\tfrac{5}{8} = 0.95 \text{ bit.}$$

c. If the colour of her hair is given as well as that of her eyes, one speaks of a joint event with four possible outcomes. Using the results of the previous sub-questions, we find for the probabilities of these outcomes that $p(\text{blond,blue}) = 3/16$, $p(\text{blond,not blue}) = 1/16$, $p(\text{not blond,blue}) = 5/16$, $p(\text{not blond,not blue}) = 7/16$.
One receives an amount of information equal to

$$H(\text{colour eyes, colour hair})$$

$$= -\tfrac{3}{16}\log\tfrac{3}{16} - \tfrac{1}{16}\log\tfrac{1}{16} - \tfrac{5}{16}\log\tfrac{5}{16} - \tfrac{7}{16}\log\tfrac{7}{16}$$

$$= 1.75 \text{ bits.}$$

1.7. *a.* The four possible situations, namely the combinations of being suitable or not and being rejected or not, can be taken as starting point. If we denote these situations by s, $\bar{s}$, r, $\bar{r}$ then the following is given:

$$p(\bar{s}) = \tfrac{1}{4}, \qquad p(r/\bar{s}) = \tfrac{3}{4}, \qquad p(r) = \tfrac{1}{2}.$$

Hence

$$p(s) = 1 - p(\bar{s}) = \tfrac{3}{4},$$

$$p(\bar{r}/\bar{s}) = 1 - \tfrac{3}{4} = \tfrac{1}{4},$$

$$p(\bar{r}) = 1 - p(r) = \tfrac{1}{2}.$$

Furthermore, according to Bayes,

$$p(\bar{s}/r) = \frac{p(r/\bar{s})\, p(\bar{s})}{p(r)} = \frac{\tfrac{3}{4}\tfrac{1}{4}}{\tfrac{1}{2}} = \tfrac{3}{8}.$$

Since $p(\bar{s}/r) + p(s/r) = 1$, it follows that

$$p(s/r) = 1 - \tfrac{3}{8} = \tfrac{5}{8}.$$

Likewise, it can be calculated that

$$p(\bar{s}/\bar{r}) = \frac{p(\bar{r}/\bar{s})\, p(\bar{s})}{p(\bar{r})} = \frac{\tfrac{1}{4}\tfrac{1}{4}}{\tfrac{1}{2}} = \tfrac{1}{8},$$

$$p(s/\bar{r}) = 1 - p(\bar{s}/\bar{r}) = \tfrac{7}{8}.$$

Finally, the combined probabilities of each of the four combinations can be determined from the general relation

$$r_{ij} = p_i q_{ji} = q_j p_{ij}.$$

Thus we find

$$p(s,r) = p(r)\, p(s/r) = \tfrac{1}{2}\tfrac{5}{8} = \tfrac{5}{16},$$

$$p(\bar{s},r) = p(r)\, p(\bar{s}/r) = \tfrac{1}{2}\tfrac{3}{8} = \tfrac{3}{16},$$

$$p(s,\overline{r}) = p(s)\,p(\overline{r}/s) = \frac{3}{4}\,\frac{7}{12} = \frac{7}{16},$$

$$p(\overline{s},\overline{r}) = p(\overline{s})\,p(\overline{r}/\overline{s}) = \frac{1}{4}\,\frac{1}{4} = \frac{1}{16}.$$

The results can be shown in a diagram, where the probabilities have been multiplied by 16. See Figure 1.11.

As only the result of the selection is mentioned without any further specification, there are two possible results of the selection. The amount of information is

H(selection/not suitable)

$$= -p(r/\overline{s})\log p(r/\overline{s}) - p(\overline{r}/\overline{s})\log p(\overline{r}/\overline{s})$$

$$= -\frac{3}{4}\log\frac{3}{4} - \frac{1}{4}\log\frac{1}{4} = 0.81 \text{ bit.}$$

b. If the selection is made by tossing a coin, each student will have a 50% chance of being rejected. As a result all of the probabilities $p(r/s)$, $p(r/\overline{s})$, $p(\overline{r}/s)$ and $p(\overline{r}/\overline{s})$ become $\frac{1}{2}$, irrespective of the condition s or $\overline{s}$. The amount of information becomes

$$H(\text{selection/not suitable}) = H(\text{selection}) = -\frac{1}{2}\log\frac{1}{2} - \frac{1}{2}\log\frac{1}{2}$$

$$= 1 \text{ bit.}$$

c. Since being suitable or not does not play a role in b, the student can make no use of his foreknowledge, namely that he is not suitable. The amount of information that he receives in b is identical to the information that is received after tossing a coin and therefore equal to 1 bit. The uncertainty is smaller for a, so that he also receives less information.

Figure 1.11. Joint probabilities of Exercise 1.7.

	r	$\overline{r}$	
s	5	7	12
$\overline{s}$	3	1	4
	8	8	

1.8. *a.* Using the given matrix, it directly follows that

$$H(X,Y) = -\sum_{i=1}^{3}\sum_{j=1}^{3} r_{ij} \log r_{ij}$$

$$= -\left(2 \times \frac{7}{24}\log\frac{7}{24} + 4 \times \frac{1}{24}\log\frac{1}{24} + \frac{1}{4}\log\frac{1}{4}\right) = 2.30 \text{ bits.}$$

b. Since for all *j*

$$q_j = \sum_{i=1}^{3} r_{ij}$$

it follows that $q(y_1) = q(y_2) = q(y_3) = \frac{1}{3}$. The amount of information then becomes

$$H(Y) = \log 3 = 1.58 \text{ bits.}$$

c. We are asked to calculate $H(X/Y)$. This is most easily calculated from $H(X,Y) = H(Y) + H(X/Y)$. This gives $H(X/Y) = 0.72$ bit. One can also determine the conditional probabilities p_{ij} from r_{ij} and q_j and substitute them in

$$H(X/Y) = -\sum_{i=1}^{3}\sum_{j=1}^{3} r_{ij} \log(p_{ij}).$$

1.9. *a.* We have $p(v_1/u_0) = 1 - p(v_0/u_0) = 1/4$.
Thus for the uncertainty with regard to the received symbol, given that a 'zero' has been transmitted, we find

$$H(V/u_0) = -p(v_0/u_0) \log p(v_0/u_0) - p(v_1/u_0) \log p(v_1/u_0)$$

$$= -\frac{3}{4}\log\frac{3}{4} - \frac{1}{4}\log\frac{1}{4} = 0.82 \text{ bit.}$$

b. First calculate the joint probabilities.
We have $p(u_0,v_0) = p(v_0/u_0)p(u_0) = 3/8$. It can similarly be found that

$$p(u_0,v_1) = \frac{1}{8}, \ p(u_1,v_0) = \frac{1}{4} \text{ and } p(u_1,v_1) = \frac{1}{4}.$$

For the amount of information with regard to the received symbol, given the transmitted symbol, it now follows that

$$H(V/U) = -\sum_{i=0}^{1}\sum_{j=0}^{1} p(u_i,v_j) \log p(v_j/u_i)$$

$$= -\frac{3}{8}\log\frac{3}{4} - \frac{1}{8}\log\frac{1}{4} - \frac{1}{4}\log\frac{1}{2} - \frac{1}{4}\log\frac{1}{2} = 0.91 \text{ bit.}$$

c. Method I: Filling in the joint probabilities in the formula for the joint information yields

$$H(U,V) = -\sum_{i=0}^{1}\sum_{j=0}^{1} p(u_i,v_j) \log p(u_i,v_j) = 1.91 \text{ bits.}$$

Method II: Since $p(u_0) = p(u_1) = \frac{1}{2}$, the amount of information $H(U)$ with regard to the transmitted symbol is equal to $H(U) = 1$ bit. It now follows that

$$H(U,V) = H(U) + H(V/U) = 1 + 0.91 = 1.91 \text{ bits.}$$

In this case, this method is faster than method I.

d. Since it can be derived that $p(v_0) = 5/8$ and $p(v_1) = 3/8$, it follows for the information $H(V)$ with regard to the received symbol that

$$H(V) = -\frac{5}{8}\log\frac{5}{8} - \frac{3}{8}\log\frac{3}{8} = 0.96 \text{ bit.}$$

Hence

$$H(U/V) = H(U,V) - H(V) = 1.91 - 0.96 = 0.95 \text{ bit.}$$

2

The discrete memoryless information source

2.1 The discrete information source

As a *discrete information source* we understand a source that generates a sequence of *symbols* (sometimes also called letters), where each symbol belongs to the same set of possible symbols. This set of possible symbols is called the *source alphabet*. The symbols are denoted by $u_1, u_2, \ldots, u_n$ and the alphabet by

$$U = \{u_1, \ldots, u_i, \ldots, u_n\}.$$

These symbols are generated at discrete points in time, which is why, together with the fact that the source alphabet is finite, we speak of a discrete information source. A group of consecutive symbols is called a *message or word.*

A certain analogy exists with the written language. The source alphabet U can then be regarded as the alphabet with its 26 letters, the space and possibly a few punctuation marks. In the written language words consist of groups of letters, separated by a space.

Words or messages of length l will be denoted by v. Since the alphabet consists of n symbols, the number of possible messages is equal to n^l. The set $V = \{v_1, \ldots, v_j, \ldots, v_{nl}\}$ is the set of all possible messages.

By making a distinction between symbols and messages we can view the information source in two ways: at the symbolic level and at the message level. We assume that the information source is stochastic, that is, that the symbols of the alphabet U each have a certain probability of occurring. Denoting these probabilities by $p(u_1) = p_1$, $p(u_2) = p_2$, etc., we are concerned with a probability distribution related to U :

$$P = \{p_1, \ldots, p_i, \ldots, p_n\}.$$

Throughout this whole book we implicitly assume that the probabilities being considered remain unchanged with the passage of time. In many applications this assumption is justified. One then says that the generated symbols form a stochastic sequence which is *stationary*.

A second aspect that is of importance is the interdependence of consecutive symbols in a message. One then speaks of the memory of the source. In this chapter we will consider *a memoryless source*, that is to say that the generated symbols are statistically independent.

Considering the information source at the symbolic level, the amount of information that is generated by the discrete memoryless source is equal to

$$H(U) = -\sum_{i=1}^{n} p_i \log p_i \quad \text{bits/symbol.} \tag{2.1}$$

The maximum amount of information that can be generated by a discrete memoryless source is

$$\max_u H(U) = \log n \quad \text{bits/symbol.} \tag{2.2}$$

As we have shown in Theorem 1.1 this maximum is achieved if all symbols have the same probability of occurring, that is, if $p_i = 1/n$ for all i. By comparing the amount of information $H(U)$ with the possible maximum we get an impression of the redundancy of the source.

Definition 2.1
The *redundancy* of a discrete memoryless information source is defined as

$$\text{red} = 1 - \frac{H(U)}{\max\limits_u H(U)} = 1 - \frac{H(U)}{\log n}, \tag{2.3}$$

where $H(U)$ is the amount of information (information content) of an information source with a source alphabet with size n.   ○

Clearly, if a source generates symbols with an equal probability of occurring, then $H(U) = \max H(U)$, so that for the redundancy it yields red = 0. In the case where the source generates just one symbol, that is, one symbol has a probability of 1 while all other probabilities are 0, then $H(U) = 0$ and thus red = 1. The value of the redundancy will therefore vary between 0 and 1.

A source that can generate two symbols is called a *binary source*. From the foregoing, it follows that such a source can generate a maximum of log 2 = 1 bit/symbol, but for the general case where the symbol probabilities are p and $(1-p)$ respectively the amount of information generated will be smaller, namely

$$H(U) = -p \log p - (1 - p)\log(1 - p) \text{ bits/symbol,}$$

and thus the redundancy will be larger.

Example 2.1
A binary memoryless source generates the symbols 0 and 1 with probabilities $\frac{1}{4}$ and $\frac{3}{4}$. The alphabet is therefore $U = \{0,1\}$, while we further have that

$$H(U) = -\frac{1}{4} \log \frac{1}{4} - \frac{3}{4} \log \frac{3}{4} = 0.81 \text{ bit/symbol}$$

and

$$\max_u H(U) = \log 2 = 1 \text{ bit/symbol.}$$

The redundancy is then equal to

$$\text{red} = 1 - \frac{H(U)}{\max\limits_u H(U)} = 1 - \frac{0.81}{1} = 0.19. \qquad \triangle$$

It will now be determined what follows for the amount of information if messages are considered instead of single symbols. First the following example will be considered.

Example 2.2
Consider the binary source from Example 2.1 and assume that messages of length 3 are being formed. As is apparent from Figure 2.1 there are now 8 possible messages, denoted by $v_1,\ldots,v_8$.

Figure 2.1. Number of possible messages of length 3 in the case of a binary source.

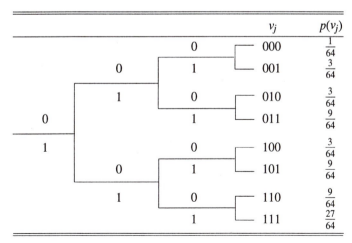

Since the source is memoryless, the symbols are statistically independent and we see that for message v_2 for example

$$p(001) = p(0)\, p(0)\, p(1) = \frac{1}{4}\frac{1}{4}\frac{3}{4} = \frac{3}{64}.$$

In other words the probability of a message is equal to the product of the probabilities of the individual symbols. The probabilities of all the messages are depicted in Figure 2.1. If the amount of information with respect to the messages is now calculated then one will find that

$$H(V) = -\frac{1}{64} \log \frac{1}{64} - 3 \times \frac{3}{64} \log \frac{3}{64} - 3 \times \frac{9}{64} \log \frac{9}{64} - \frac{27}{64} \log \frac{27}{64}$$
$$= 2.45 \text{ bits/message.}$$

Note that

$$H(V) = 3\, H(U). \hspace{4cm} \triangle$$

In general, n^l different messages of length l can be formed from a source alphabet U of size n. Since we are dealing with a memoryless source the probability of occurrence of each message is equal to the product of the probabilities of the l individual symbols of which the message is made up. The amount of information at the message level is given by

$$H(V) = -\sum_{j=1}^{n^l} p(v_j) \log p(v_j). \hspace{3cm} (2.4)$$

By writing out the probabilities $p(v_j)$ in terms of $p(u_1),\ldots,p(u_n)$ it can easily be proven that

$$H(V) = l\, H(U). \hspace{4cm} (2.5)$$

A message of length l therefore contains l times as much information as a message of length 1. Note that this is so because successive symbols are independent. We will return to the case for which this is not true later.

The amount of information of an information source can also be expressed in bits per unit of time, for example bits/second. One then speaks of the *production* of the source. If all symbols have an equal duration, say t seconds, then the production $H_t(U)$ of a source is

$$H_t(U) = \frac{1}{t} H(U) \hspace{1cm} \text{bits/second.} \hspace{2cm} (2.6)$$

If the symbols do not have equal duration, as is the case with Morse code for example where the "dash" takes longer than the "dot", the average amount of time t is used.

2.2 Source coding

Consider a memoryless information source that generates messages made up from the symbols from the source alphabet $U = \{u_1, u_2, \ldots, u_n\}$. According to the communication model of Figure 1.9, in general, first, *data reduction* is applied in order to remove information that is not relevant for the destination. Here, we assume that the information source generates messages which cannot be reduced further.

For reasons of efficiency we shall want to represent the messages as compactly as possible by removing the redundancy present within the message. This process is known as *source coding*. We will limit ourselves to but a few examples of codes that can be used for encoding messages. Since the information source is assumed to be memoryless, it suffices to consider the encoding of the separate source symbols instead of the messages.

Let the code alphabet be given by $S = \{s_1, s_2, \ldots, s_r\}$. We will now look for codes that give a certain combination of code symbols for every source output symbol, a so-called *code word*.

If all code words are different we speak of a *non-singular code*. If for a succession of code words the result also remains non-singular, one has a code that is *uniquely decodable*. In the case of unique decodability the received message must have a single unique possible interpretation. Finally, in the case where a code is uniquely decodable, and each message symbol can also be directly decoded, that is without first looking at succeeding code symbols, then one speaks of an *instantaneous code*.

Example 2.3
An information source has an alphabet with four source output symbols u_1, u_2, u_3 and u_4. The code alphabet consists of the two symbols 0 and 1. Code words are made up using four different coding systems according to the following table.

	A	B	C	D
u_1	0	00	0	0
u_2	11	01	10	01
u_3	00	10	110	011
u_4	01	11	1110	0111

The four codes are all non-singular. Code A is not uniquely decodable, however, because the combination 0011 for example can be obtained from $u_1 u_1 u_2$ or from $u_3 u_2$. The codes B, C and D are uniquely decodable: B because all code words are equally long, so that one need only split the consecutive sequence of code words into groups of two symbols; C because every code word ends with a 0, which thus functions as a comma ("comma code"), D because every code word begins with a 0. Code D is not instantaneously decodable, because one must always wait for the first symbol of the next code word before the current code word can be decoded.

$$\triangle$$

It is a necessary and sufficient requirement for an instantaneous code that a complete code word never forms the beginning of another code word. In this case one also speaks of a *prefix code*.

From the various limitations indicated above, it follows that there exists a one-to-one relation between the source output symbols with the alphabet $U = \{u_1, u_2, \dots, u_n\}$ and the code words. For simplicity, we will therefore also denote these code words by $u_1, u_2, \dots, u_n$. We will denote the lengths of these code words by $l_1, l_2, \dots, l_n$. These lengths are determined by the number of code symbols making up the code word.

In the following theorem we examine what requirements must be met by a code in order for it to be instantaneously decodable.

Theorem 2.1 (Kraft's inequality)
It is a necessary and sufficient requirement for the existence of an instantaneous code that

$$\sum_{i=1}^{n} r^{-l_i} \leq 1, \tag{2.7}$$

where r is the size of the code alphabet, and l_i, $i = 1, \dots, n$, the length of the code word u_i.

Proof
Assume that the number of code words of length 1 is equal to w_1. This number will at most be equal to r ($w_1 \leq r$). The code words being used may not form the beginning of another code word. Thus $r - w_1$ first code symbols remain. For the number of code words of length 2 we thus have

$$w_2 \leq (r - w_1)r = r^2 - w_1 r.$$

Similarly $w_3 \leq \{(r - w_1)\, r - w_2\} r = r^3 - w_1 r^2 - w_2 r.$

And if m is the maximum length of the code words, then

$$w_m \leq r^m - w_1 r^{m-1} - w_2 r^{m-2} - \ldots - w_{m-1} r.$$

Dividing by r^m gives

$$0 \leq 1 - w_1 r^{-1} - w_2 r^{-2} - \ldots - w_{m-1} r^{-m+1} - w_m r^{-m}.$$

Or

$$\sum_{j=1}^{m} w_j r^{-j} \leq 1.$$

That is

$$\underbrace{\frac{1}{r} + \frac{1}{r} + \ldots + \frac{1}{r}}_{w_1} + \underbrace{\frac{1}{r^2} + \frac{1}{r^2} + \ldots + \frac{1}{r^2}}_{w_2} + \ldots + \underbrace{\frac{1}{r^m} + \frac{1}{r^m} + \ldots + \frac{1}{r^m}}_{w_m} \leq 1.$$

But $w_1 + w_2 + \ldots + w_m = n$, namely the total number of code words, so that this inequality is identical to

$$\sum_{i=1}^{n} r^{-l_i} \leq 1. \qquad \square$$

Note that Kraft's inequality indicates that an instantaneous code exists that has code words of these lengths l_i. This does not mean, however, that every code that satisfies the inequality is instantaneous.

The next step concerns the manner in which one makes an appropriate choice for the lengths l_i. One would like the lengths of the code words to depend on the probability of occurrence of the message in order to guarantee an optimal use of the channel. That is, one will prefer to give messages with a large probability of occurring a shorter code word than messages with a small probability of occurring.

Example 2.4
With Morse code, where letters from the alphabet are converted into code words composed of dots and dashes, code words for frequently appearing letters (such as the letter e for example) have been chosen that consist of as few dots and dashes as possible on the one hand, and consists preferably of dots on the other hand since these have a shorter duration than dashes. A dot spends 2 units of times, whereas a dash needs 4 units of time. A letter space consists of 3 time units. In Figure 2.2 the Morse code is given. $\triangle$

The following theorem gives the relationship between the average code word length L and the amount of information of an information source.

Theorem 2.2 (Source coding theorem)
Consider a set of n code words u_i with probability distribution $P = (p_1,\ldots,p_n)$, $p_i > 0$ for all i, where all code words are composed of symbols from the code alphabet $S = (s_1,s_2,\ldots,s_r)$. If Kraft's inequality is satisfied then

$$\frac{H(U)}{\log r} \le L, \tag{2.8}$$

where L is the average code word length, defined by

$$L = \sum_{i=1}^{n} p_i l_i, \tag{2.9}$$

and l_i is the length of code word u_i.
Equality holds if and only if $p_i = r^{-l_i}$ for $i = 1,\ldots,n$.

Proof
We have

$$H(U) - L \log r = -\sum_{i=1}^{n} [p_i \log p_i + p_i l_i \log r]$$

Figure 2.2. The Morse code (probabilities are for English language).

Symbol	Probability	Morse code	Symbol	Probability	Morse code
A	0.0642	· —	N	0.0574	— ·
B	0.0127	— · ·	O	0.0632	— — —
C	0.0218	— · — ·	P	0.0152	· — — ·
D	0.0317	— · ·	Q	0.0008	— — · —
E	0.1031	·	R	0.0484	· — ·
F	0.0208	· · — ·	S	0.0515	· · ·
G	0.0152	— — ·	T	0.0796	—
H	0.0467	· · · ·	U	0.0228	· · —
I	0.0575	· ·	V	0.0083	· · · —
J	0.0008	· — — —	W	0.0175	· — —
K	0.0049	— · —	X	0.0014	— · · —
L	0.0321	· — · ·	Y	0.0164	— · — —
M	0.0198	— —	Z	0.0005	— — · ·
			Space	0.1859	

$$= \sum_{i=1}^{n} p_i \log \left\{ \frac{1}{p_i r^{l_i}} \right\} = \sum_{i=1}^{n} p_i \frac{\ln \left\{ \frac{1}{p_i r^{l_i}} \right\}}{\ln 2}. \qquad (2.10)$$

Since for $a > 0$

$$\ln a \leq a - 1$$

with equality if $a = 1$ (see Figure 1.2), it follows from equation (2.10) that

$$\sum_{i=1}^{n} p_i \ln \left(\frac{1}{p_i r^{l_i}} \right) \leq \sum_{i=1}^{n} p_i \left(\frac{1}{p_i r^{l_i}} - 1 \right) = \sum_{i=1}^{n} r^{-l_i} - 1.$$

Since the Kraft inequality is satisfied it consequently follows that

$$H(U) - L \log r \leq 0,$$

which yields exactly formula (2.8). The requirement for equality can be directly derived from this. □

What Theorem 2.2 actually says is that the average length can never be smaller than the amount of information (with base r) of an information source. The smallest average length of code words arises when the lengths l_i are chosen in such a way that the equals sign in equation (2.8) holds. This is so if for all i

$$p_i = r^{-l_i} \quad \text{or also} \quad l_i = -\log_r p_i. \qquad (2.11)$$

This can only be achieved if $-\log_r p_i$ is a whole number, since l_i is the length of a code word made up from a whole number of symbols each of length 1. If $-\log_r p_i$ is not a whole number, no optimal coding will be possible. It then seems obvious to choose the integer lying directly above as the length l_i. One thus chooses l_i in such a way that

$$-\log_r p_i \leq l_i < -\log_r p_i + 1. \qquad (2.12)$$

Note that in the case of an optimal code, i.e. $l_i = -\log_r p_i$, we have that $\sum_i r^{-l_i} = 1$ (compare with Kraft's inequality) and in the other cases $\sum_i r^{-l_i} < 1$.

In view of Theorem 2.2, it is now also a simple matter to construct a criterion for the quality of a code. The closer $H(U) / (L \log r)$ approaches the value 1 (i.e. the equals sign in equation (2.8) holds), the more *efficient* the code.

Definition 2.2

The *efficiency* η of a code is defined as

$$\eta = \frac{H(U)}{L \log r},$$ (2.13)

where $H(U)$ is the amount of information of the source, L is the average length of the code word and r is the size of the code alphabet.   ○

Example 2.5

An information source has an alphabet with four source output symbols u_1, u_2, u_3 and u_4. The code alphabet consists of the two symbols 0 and 1. The probabilities of the message symbols are all 1/4. Assume the code is as follows:

symbol	code ($r = 2$)
u_1	00
u_2	01
u_3	10
u_4	11

For this code $H(U) = \log 4 = 2$ bits, $r = 2$ and $L = 2$. Thus the efficiency of this code is

$$\eta = \frac{H(U)}{L \log r} = \frac{2}{2 \times 1} = 1 = 100\%.$$ △

Example 2.6

For the same source, the probabilities of the source output symbols are now $\frac{1}{2}, \frac{1}{4}, \frac{1}{8}$ and $\frac{1}{8}$. For the same code we now have

$$H(U) = -\frac{1}{2} \log \frac{1}{2} - \frac{1}{4} \log \frac{1}{4} - 2 \times \frac{1}{8} \log \frac{1}{8} = \frac{7}{4} \text{ bits, } r = 2, L = 2.$$

The efficiency is now

$$\eta = \frac{7/4}{2 \times 1} = \frac{7}{8} = 87.5\%.$$ △

Example 2.7

The same information source now uses the code shown in the table.

We again find that $H(U) = 7/4$ and $r = 2$. But

symbol	code ($r = 2$)
u_1	0
u_2	10
u_3	110
u_4	111

now

$$L = p_1 \, 1 + p_2 \, 2 + p_3 \, 3 + p_4 \, 3 = \frac{1}{2} \, 1 + \frac{1}{4} \, 2 + \frac{1}{8} \, 3 + \frac{1}{8} \, 3 = \frac{7}{4}.$$

And the efficiency is therefore

$$\eta = \frac{7/4}{7/4 \times 1} = 1 = 100\%.$$

Note that the requirement for an optimal code is indeed satisfied: $p_i = r^{-l_i}$ for all i. △

2.3 Coding strategies

In the previous section the requirements for the existence of optimal codes were examined, that is to say codes with a high efficiency. Various strategies are known to find a code that approaches this optimum. For the first three codes which will be given hereafter the source output symbols are placed in order of decreasing probability.

I Fano code
After having placed the source output symbols in order of decreasing probability, Fano divides the symbols as well as possible into r equally probable groups. Each group receives one of the r code symbols as the first symbol. This division is repeated per group as many times as this is possible.

Example 2.8

symbol	probability	binary code ($r = 2$)
u_1	1/4	0<u>0</u>
u_2	1/4	0<u>1</u>
u_3	1/8	10<u>0</u>
u_4	1/8	1<u>01</u>
u_5	1/16	110<u>0</u>
u_6	1/16	11<u>01</u>
u_7	1/32	1110<u>0</u>
u_8	1/32	111<u>01</u>
u_9	1/32	1111<u>0</u>
u_{10}	1/32	11111

△

In Example 2.8 it is the case that $H(U) = L = 2.88$ and thus the efficiency of the code is equal to $\eta = 1$. However, it is not always possible to separate precisely the probabilities into equally probable groups. In that case the division should be as good as possible. The result can be various alternatives.

Example 2.9

symbol	probability	code 1 $(r = 1)$	code 2 $(r = 2)$
u_1	1/3	0<u>0</u>	<u>0</u>
u_2	1/3	0<u>1</u>	1<u>0</u>
u_3	1/9	1<u>0</u>	11<u>0</u>
u_4	1/9	11<u>0</u>	111<u>0</u>
u_5	1/9	111	1111

△

In Example 2.9 the average code word lengths will be the same ($L = 2.22$). If this is not the case the code with the best efficiency is preferable as a matter of fact. Fano coding is also applicable in situations where $r > 2$.
Due to the source coding theorem (compare theorem 2.2), in general the average code word length will decrease for increasing size r of the alphabet. See Example 2.10.

Example 2.10

symbol	probability	code 1 $(r = 2)$	code 2 $(r = 3)$	code 3 $(r = 4)$
u_1	0.30	0<u>0</u>	<u>0</u>	<u>0</u>
u_2	0.25	0<u>1</u>	10	<u>1</u>
u_3	0.12	10<u>0</u>	1<u>1</u>	2<u>0</u>
u_4	0.10	10<u>1</u>	2<u>0</u>	2<u>1</u>
u_5	0.10	11<u>0</u>	2<u>1</u>	3<u>0</u>
u_6	0.05	111<u>0</u>	22<u>0</u>	3<u>1</u>
u_7	0.04	1111<u>0</u>	22<u>1</u>	3<u>2</u>
u_8	0.04	11111	222	33

$H(U) = 2.64$ bits/symbol	$L_1 = 2.66$	$L_2 = 1.83$	$L_3 = 1.45$
	$\eta_1 = 0.99$	$\eta_2 = 0.91$	$\eta_3 = 0.91$

△

II Shannon code

Shannon calculates a series of cumulative probabilities $P_k = \sum\limits_{i=1}^{k-1} p(u_i)$ for $k = 1,2,\ldots,n$. These are subsequently (for a binary code) written in binary notation. The number of symbols in each code word is found from the inequality

$$\log \frac{1}{p_k} \leq l_k < \log \frac{1}{p_k} + 1. \tag{2.14}$$

In the following example the source output symbol u_6 is represented by 1101 since $P_6 = 13/16$, which can be written as $1 \cdot 2^{-1} + 1 \cdot 2^{-2} + 0 \cdot 2^{-3} + 1 \cdot 2^{-4}$. The number of code symbols must be at least $\log(1/p_k) = \log 16 = 4$, so that no zeros need to be added to the code for this symbol.

Example 2.11

symbol	probability	P_i		length l_i		code ($r = 2$)
u_1	1/4	P_1	$= 0$	l_1	$= 2$	$= 00$
u_2	1/4	P_2	$= 1/4$	l_2	$= 2$	$= 01$
u_3	1/8	P_3	$= 1/2$	l_3	$= 3$	$= 100$
u_4	1/8	P_4	$= 5/8$	l_4	$= 3$	$= 101$
u_5	1/16	P_5	$= 3/4$	l_5	$= 4$	$= 1100$
u_6	1/16	P_6	$= 13/16$	l_6	$= 4$	$= 1101$
u_7	1/32	P_7	$= 7/8$	l_7	$= 5$	$= 11100$
u_8	1/32	P_8	$= 29/32$	l_8	$= 5$	$= 11101$
u_9	1/32	P_9	$= 15/16$	l_9	$= 5$	$= 11110$
u_{10}	1/32	P_{10}	$= 31/32$	l_{10}	$= 5$	$= 11111$

$\triangle$

It may be concluded from Example 2.11 that in this case Shannon's method leads to the same code as the method of Fano (compare Example 2.8). That this is not always the case is shown by the following example.

Example 2.12

symbol	probability P_i	l_i		Shannon code ($r = 2$)	Fano code ($r = 2$)
u_1	0.4	0	2	00	0
u_2	0.3	0.4	2	01	10
u_3	0.2	0.7	3	101	110
u_4	0.1	0.9	4	1110	111

$\triangle$

The particular result of Fano coding in Example 2.12 is also called a *comma code*, since the binary digit 0 indicates the end of a code word plus the fact that no code word has a length larger than 3.

III Huffman code

With the Huffman code in the binary case the two least probable source output symbols are joined together, resulting in a new message alphabet with one less symbol. These are then arranged anew, after which two symbols are again joined together. This is carried on until a message alphabet of just two symbols arises. These two symbols are assigned the symbols 0 and 1 from the binary code. Working backwards, a 0 or a 1 is added to the code word at each place where two symbols have been joined together.

Example 2.13

symbol	probability					code ($r = 2$)
u_1	0.4	0.4	0.4	0.4	0.6 (0)	1
u_2	0.3	0.3	0.3	0.3 (0)	0.4 (1)	00
u_3	0.1	0.1	0.2 (0)	0.3 (1)		011
u_4	0.1	0.1 (0)	0.1 (1)			0100
u_5	0.06 (0)	0.1 (1)				01010
u_6	0.04 (1)					01011

$\triangle$

The example given would give a result comparable to that with Fano's method; the method of Shannon gives a less efficient code. In general the method of Huffman leads to a more efficient code.

If the number of code symbols is r, then for an optimal Huffman code there must be $r + k(r - 1)$ source output symbols (k integer). If one has fewer source output symbols then one must add symbols, and set their probabilities equal to 0. Consider the following example.

Example 2.14

symbol	probability			code ($r = 3$)
u_1	1/3	1/3	4/9 (0)	1
u_2	1/6	2/9	1/3 (1)	00
u_3	1/6	1/6 (0)	2/9 (2)	01
u_4	1/9	1/6 (1)		02
u_5	1/9 (0)	1/9 (2)		20
u_6	1/9 (1)			21
u_7	0 (2)			22

For the average code word length L it is the case that $L = 1.67$. If symbol u_7 was not added and the Huffman code was applied, then the result would be $L = 2$. $\triangle$

IV Gilbert–Moore code (alphabetic code)
A completely different method is followed by the Gilbert–Moore code. With this method the source output symbols may be placed in any desired order (e.g. alphabetic). If the length of a code word u_i is given by l_i, then this length is determined by

$$2^{1-l_i} \leq p(u_i) < 2^{2-l_i}, \quad i = 1,2,\ldots,n. \tag{2.15}$$

The non-decreasing series $(\alpha_1, \alpha_2,\ldots)$ is subsequently determined according to

$$\left. \begin{aligned}
\alpha_1 &= \tfrac{1}{2} p(u_1), \\
\alpha_2 &= p(u_1) + \tfrac{1}{2} p(u_2), \\
&\vdots \\
\alpha_i &= p(u_1) + p(u_2) + \ldots + p(u_{i-1}) + \tfrac{1}{2} p(u_i).
\end{aligned} \right\} \tag{2.16}$$

We have here that $0 \leq \alpha_1 \leq \alpha_2 \ldots \leq 1$. The code for message u_i is given by representing the number α_i in a binary series to a length of l_i.

Example 2.15
Take the first three letters of the alphabet a, b and c. With the given probabilities of occurring for the English language (see Figure 2.2) we find:

symbol	probability	l_i	α_i	code ($r = 2$)	
u_1	0.064	5	0.032	00001	
u_2	0.013	8	0.071	00010010	
u_3	0.022	7	0.088	00010111	△

V The arithmetic code
With this code the probabilities $p(u_1)$, $p(u_2),\ldots,p(u_n)$ of the source output

Figure 2.3. Coding of the sequence $u_1,u_2,\ldots$ with the help of an arithmetic code.

	u_1		u_2	u_3
0			0.5	0.8
u_1	u_2	u_3		
0.0	0.25	0.4	0.5	
	u_1	u_2	u_3	
	0.25	0.325	0.37	0.40

symbols are depicted as sub-intervals of the unit interval [0,1] (the sum of the probabilities is one). We will treat the essence of the method for the case that $n = 3$, with $p(u_1) = 0.5$, $p(u_2) = 0.3$ and $p(u_3) = 0.2$. The cumulative probabilities are $P_1 = 0$, $P_2 = 0.5$, $P_3 = 0.8$ respectively. This is depicted in Figure 2.3. Each point represents the cumulative probability P_i of the symbols $u_1, u_2, \ldots, u_{i-1}$.

If source output symbol u_1 has occurred (with probability 0.5) this corresponds with the interval [0.0, 0.5]. We subsequently assume that a second symbol is generated by the source. The current interval, namely [0.0; 0.5] is now divided up into sub-intervals again according to the cumulative probability distribution of the source, namely 0.0, 0.5, 0.8, 1.0. One now considers the three sub-intervals [0.0, 0.25], [0.25, 0.40] and [0.40, 0.50]. If symbol u_2 is now generated then one finds oneself, after two source symbols, in the interval [0.25, 0.40]. By repeating this process we produce the result that a sequence of source symbols is represented by a sub-interval of [0,1] that is uniquely coupled to it. We now take the binary representation of the leftmost point of the interval as the code word belonging to the sequence of source symbols $u_1, u_2, \ldots$ etc. The width of the interval corresponds with the probability that the corresponding sequence of source symbols occurs.

We can now regard the code as a recursive process that proceeds as follows. For each step, that is, when a new source symbol is presented again, we consider the leftmost point C of the current interval and the current width A of this interval. The new leftmost point is the old leftmost point plus a part of the current interval width, according to

$$C_{\text{new}} = C_{\text{old}} + A_{\text{old}} P_i, \qquad (2.17)$$

where P_i is the cumulative probability for symbol u_i. The width of the new interval arises by multiplying the old width by the probability p_i, thus

$$A_{\text{new}} = A_{\text{old}} p_i. \qquad (2.18)$$

In the example of Figure 2.3 we have in the beginning (zeroth step)

$$C_{\text{start}} = 0.0,$$

$$A_{\text{start}} = 1.0.$$

After u_1, the following result for C and A:

$$C_{\text{new}} = 0.0 + 1.0 \times 0.0 = 0.0,$$

$A_{\text{new}} = 1.0 \times 0.5 = 0.5.$

After the second symbol, thus u_2 in the example, we subsequently find

$C_{\text{new}} = 0.0 + 0.5 \times 0.5 = 0.25,$

$A_{\text{new}} = 0.5 \times 0.3 = 0.15.$

If we constantly encode a group of l symbols together according to this recursive path we find a left point C and the width A of the resulting interval. In order to find the final code word we choose the left point C and write it in binary form with a number of bits in such a way that it can be distinguished from the left points of the other intervals. Thus in the example the pair of source symbols $u_1 u_2$ is encoded as the binary representations of 0.25 decimal and as 0.01 binary. The decoding actually follows the inverse process by determining, step by step, through which symbol the current interval has arisen, and determining the preceding interval from this.

VI Coding on the basis of alphabet extension
Although the source is considered to be memoryless in this chapter and successive source symbols are thus independent, the coding of combinations of symbols can nevertheless lead to an efficient code. We then group l source symbols into a message, calculate the probability of these messages and subsequently use a coding strategy such as that of Huffman to obtain the eventual code. We will discuss this method, which is known as alphabet extension, with the help of an example.

Example 2.16
Two source output symbols u_1 and u_2 have probabilities of $\frac{3}{4}$ and $\frac{1}{4}$. Coding has taken place on the basis of Fano's coding method.

symbol	probability	code ($r = 2$)
u_1	3/4	0
u_2	1/4	1

with $H(U) = 0.811$ bit, $L = 1$, $r = 2$, $\eta = 0.811$.

We subsequently take two symbols together, and thus get new messages $v_1, \ldots, v_4$.

We now have

$H(V) = 1.622$, $L = 27/16$, $r = 2$, $\eta = 0.961$.

message	probability	code ($r = 2$)
$v_1 = u_1 u_1$	$p(v_1) = p(u_1, u_2) = 9/16$	0
$v_2 = u_1 u_2$	$p(v_2) = p(u_1, u_2) = 3/16$	10
$v_3 = u_2 u_1$	$p(v_3) = p(u_2, u_1) = 3/16$	110
$v_4 = u_2 u_2$	$p(v_4) = p(u_2, u_2) = 1/16$	111

The efficiency has thus been increased by taking two source symbols together. $\triangle$

By encoding l source symbols together a new source results with an extended alphabet, namely with n^l messages (instead of the original one consisting of n symbols), where the probability of a message is the product of the probabilities of the symbols which make up the message.

2.4 Most probable messages

In Section 2.1 we introduced the set $V = \{v_1, v_2, \ldots, v_j, \ldots, v_{nl}\}$ with messages v consisting of l source symbols from a source alphabet with size n. For increasing l it appears that some messages have a negligible probability of occurrence while the others have approximately the same probability. Then we can speak of the *number of most probable messages* which is, of course, smaller than the number of possible messages.

Let ℓ_i ($\ell_i > 0$) be the number of times that source symbol u_i appears in a message v_j. Then in the case of a memoryless source it is the case that the probability for an arbitrary message v is given by

$$p(v) = \prod_{i=1}^{k} p(u_i)^{\ell_i}, \tag{2.19}$$

where k is the number of different source symbols in the message v and

$$\ell = \sum_{i=1}^{k} \ell_i. \tag{2.20}$$

And thus

$$\log p(v) = \log \left\{ \prod_{i=1}^{k} p(u_i)^{\ell_i} \right\}. \tag{2.21}$$

If by the law of large numbers $\ell_i / \ell \to p(u_i)$, i.e. $\ell_i \approx \ell p(u_i)$, we find

$$\log p(v) \approx \log \left\{ \prod_{i=1}^{k} p(u_i)^{\ell p(u_i)} \right\}$$

$$= \ell \sum_{i=1}^{k} p(u_i) \log p(u_i)$$

$$= -\ell \, H(U). \tag{2.22}$$

Thus

$$\frac{1}{\ell} \log \, p(v) \approx -H(U). \tag{2.23}$$

Thus $H(U)$ is approximately the logarithm of the reciprocal probability of a typical long sequence divided by the number of symbols in a sequence. This holds for any source. More precisely we have the following theorem.

Theorem 2.3 (Shannon–McMillan theorem)
Given a discrete memoryless source with alphabet U and information $H(U)$. Then given any $\varepsilon > 0$ and $\delta > 0$, we can find an ℓ_0 such that the source words of any length $\ell \geq \ell_0$ fall into two classes.
(i)   A set S' whose total probability is less than ε.
(ii) The remainder, the set S, all of whose members have probabilities satisfying the inequality

$$\left| \frac{-\log p(v)}{\ell} - H(U) \right| \leq \delta. \tag{2.24}$$

Proof
Set S can be defined as

$$S = \{ v \mid -\log p(v) - \ell H(U) | < \ell\delta \}. \tag{2.25}$$

Chebyshev's inequality states that for any random variable x with mean μ and variance σ^2

$$p\{ \, |x - \mu| \geq \varepsilon \} \leq \frac{\sigma^2}{\varepsilon^2}. \tag{2.26}$$

Applying this to the set S yields

$$p\{ \, |-\log p(v) - \ell H(U)| \geq \ell\delta \} \leq \frac{\mathrm{var}[-\log p(v)]}{\ell^2 \delta^2}$$

$$\leq \frac{\ell\sigma^2}{\ell^2 \delta^2} = \frac{\sigma^2}{\ell\delta^2}, \tag{2.27}$$

where

$$\sigma^2 = \sum_{p(u_i)} p(u_i) \, (\log p(u_i))^2 - \left(\sum_{p(u_i)} p(u_i) \log p(u_i) \right)^2$$

is a constant, independent of ℓ.

Consequently, for sufficiently large ℓ, the probability $p(S')$ that source words occur belonging to set S' is less than ε.   □

Clearly, for the probability $p(S) = \sum_{v \in S} p(v)$ that source words occur belonging to the set S the following holds:

$$1 - \varepsilon < p(S) \leq 1. \tag{2.28}$$

For small ε, S contains all the source words with a high probability of occurrence. For that very reason S is also called the set of *most probable source words* or *most probable messages*. The following theorem gives boundaries for the number of elements of S, denoted by $M = |S|$.

Theorem 2.4

For the number M of source words in the set S of most probable source words it is the case that

$$(1 - \varepsilon)2^{\ell\{H(U)-\delta\}} \leq M \leq 2^{\ell\{H(U)+\delta\}}. \tag{2.29}$$

Proof
From

$$\left| \frac{-\log p(v)}{n} - H(U) \right| < \delta$$

it follows that

$$2^{-\ell\{H(U)+\delta\}} \leq p(v) \leq 2^{-\ell\{H(U)-\delta\}} \tag{2.30}$$

and thus

$$\sum_{v \in S} 2^{-\ell\{H(U)+\delta\}} < p(S) \leq \sum_{v \in S} 2^{-\ell\{H(U)-\delta\}}, \tag{2.31}$$

or

$$M \, 2^{-\ell\{H(U)+\delta\}} < p(S) \leq M \, 2^{-\ell\{H(U)-\delta\}}. \tag{2.32}$$

Also (compare equation (2.28))

$$1 - \varepsilon \leq p(S) \leq 1.$$

Thus

$$1 - \varepsilon \leq M \, 2^{-\ell\{H(U)-\delta\}}, \tag{2.33}$$

and

$$M \, 2^{-\ell\{H(U)+\delta\}} \leq 1. \tag{2.34}$$

Which proves the theorem. □

For small ε and δ it follows that

$$M \approx 2^{\ell H(u)}. \tag{2.35}$$

Because the amount of information of the source is at most equal to log n, namely if all of the source symbols have the same probability, it follows that

$$M_{\max} = 2^{\ell \log n} = n^{\ell}, \tag{2.36}$$

which is exactly the number of possible messages. For a source where $H(U)$ < log n, the number of most probable messages of length ℓ will then be smaller than the number of possible messages. One can also say that a part of the possible messages have a negligibly small probability of occurring if the length ℓ is large.

As expressed in equation (2.35) there is an exponential relationship between the number of most probable messages of a discrete information source M and the information content of the source. For a source which generates a maximum amount of information, the number of probable messages is equal to the number of possible messages. We can thus regard the amount of information as a measure for the number of messages that a source can actually generate (that is, with a certain probability).

With the help of the approach introduced here, we may further explain the material which was treated in Sections 2.2 and 2.3. We will do this here to illustrate how the concept "number of most probable messages" may be used. We assume that the source messages are of length ℓ and assign a code word of length L to each source message. The source symbols are chosen from an alphabet $U = \{u_1, u_2, \ldots, u_n\}$ and the code words from an alphabet $S = \{s_1, s_2, \ldots, s_r\}$.

The number of possible source messages is then n^{ℓ} and the number of possible code words available is r^L. It is therefore possible to encode each message into a code word if

$$r^L \geq n^{\ell}$$

or

$$\frac{L}{\ell} \geq \frac{\log n}{\log r} = \log_r n,$$

since a code word is then available for each source message. It turns out, however, that we only have to take the number of most probable messages *M* into account. This leads to the following source encoding theorem.

Theorem 2.5 (Shannon's first coding theorem)
Given a discrete information source without memory and with an amount of information $H(U)$ from which messages of length ℓ are encoded into code words of length L, from a code alphabet of size r. If P_e is the probability that a message occurs for which there is no code word available, then P_e can be made arbitrarily small, i.e. $P_e < \varepsilon$, as long as the length L satisfies

$$L \log r \geq \ell\, H(U)$$

and ℓ is sufficiently large.

Proof
From the foregoing theorem we know that the number of source words in S is less than $2^{\ell\{H(U)+\delta\}}$. The theorem assumes that

$$L \log r \geq \ell\, H(U). \tag{2.37}$$

So a δ can be chosen such that

$$L \log r \geq \ell\{H(U) + \delta\}, \tag{2.38}$$

and

$$r^L \geq 2^{\ell\{H(U)+\delta\}}. \tag{2.39}$$

The conclusion is that the number of code words of block length L ($= r^L$) is larger than the number of source words in S. As a consequence for each source word of S there is a code word.
The set for which a code word is not present is contained in S' and for ℓ large enough, $P_e < \varepsilon$ as was to be proved.   □

This means that one can encode error-free when a large length of source words is chosen, even if the number of possible code words is smaller then the number of possible messages. In fact, this is why joining the source symbols, as mentioned in Section 2.3, can lead to a more efficient code.

2.5 Exercises

In the following exercises it is assumed that consecutive symbols are statistically independent.

2.1. An information source generates symbols belonging to the alphabet $U = \{u_1,u_2,u_3\}$. The probabilities of these symbols are 0.7, 0.2, and 0.1 respectively.
a. Calculate the amount of information per symbol.
b. Calculate the probabilities of all possible messages consisting of two symbols.
c. Calculate the amount of information, for both *a* and *b*, per message of two symbols.
d. Calculate the redundancy of the information source.
e. Calculate the amount of information per second, if it is given that the duration of u_1 is 0.001 sec, of u_2 is 0.002 sec and of u_3 is 0.003 sec.

2.2. An information source produces 8 different symbols (u_1 to u_8) with respective probabilities of 1/2, 1/4, 1/8, 1/16, 1/32, 1/64, 1/128, 1/128. These symbols are encoded as 000, 001, 010, 011, 100, 101, 110, 111 respectively.
a. What is the amount of information per symbol?
b. What are the probabilities of occurring for a 0 and a 1?
c. What is the efficiency of this code?
d. Give an efficient code with the help of the method of Fano or Shannon.
e. What is the efficiency of the code so obtained?

2.3. Six message symbols ($u_1,...,u_6$) occurring with probabilities of 3/8, 1/6, 1/8, 1/8, 1/8 and 1/12 are to be encoded using a ternary code (code symbols 0, 1, 2).
a. Determine the required code by applying the method of Fano.
b. Determine the efficiency of the code obtained.

2.4. Seven message symbols ($u_1,...,u_7$) occurring with probabilities of 1/4, 1/5, 1/6, 1/6, 1/12, 1/12 and 1/20 need to be encoded using four code symbols a, b, c, d.
a. Determine the required code using the method of Fano.
b. Determine the efficiency of the code obtained.

2.5. An information source has an alphabet consisting of five symbols ($u_1,...,u_5$). The probabilities of these symbols occurring are 1/2, 1/4, 1/8, 1/16 and 1/16 respectively.
a. Determine a suitable code for these message symbols with 3 code symbols a, b and c.
b. Determine the efficiency of the code obtained.

2.6. An information source has an alphabet consisting of three symbols (a, b, c). The probabilities of these symbols occurring are 1/2, 1/3, 1/6 respectively.

a. Determine a suitable binary code for these message symbols.
b. Determine the efficiency of the code obtained.
c. An improvement of the efficiency is sought by constantly taking two message symbols together. Determine a suitable binary code for this.
d. Determine the efficiency of the code obtained in c.

2.7. An information source has a source alphabet of 9 different symbols u_1 to u_9, with respective probabilities of 1/4, 1/4, 1/8, 1/8, 1/16, 1/16, 1/16, 1/32, 1/32. This information source is linked to a communication channel, which makes use of the three symbols a, b and c.
a. Determine the code and the efficiency of the code on the basis of the methods of Fano and Huffman.
b. Answer the same questions, in the case when the symbol c may never be followed by another c.
Note that the code must be instantaneously decodable.

2.8. Two independent questions are asked in an opinion poll, which can each be answered with yes, no, no opinion. For a previous poll, where the same questions were asked, the results were as follows:

	yes	no	no opinion
question 1	50%	40%	10%
question 2	60%	20%	20%

On the basis of this data, one wants to create a code for the combination of the two answers, where one wants to make a comparison between a binary and a ternary code.
a. Determine a suitable binary code according to the method of Fano.
b. Determine a suitable ternary code according to the method of Fano.
c. Give your preference, based on the efficiency of both codes.

2.9. Give a code according to the method of Gilbert and Moore for the message symbols u_i indicated below.

u_i	u_1	u_2	u_3	u_4
$p(u_i)$	0.1	0.3	0.2	0.4

Compare this method, on the basis of the efficiency, with a coding according to Shannon and one according to Huffman.

2.10. An experiment has the numbers 1, 2, 3, 4, 5, 6 and 7 as possible outcomes with probabilities $p(1) = p(2) = 1/3, p(3) = p(4) = 1/9$ and $p(5) = p(6) = p(7) = 1/27$. One wishes to transmit the outcomes of the experiments over a binary or a ternary channel. Both channels are noise-free. The costs of the binary channel are £ 1.80 per code symbol and £ 2.70 per code symbol for the ternary channel.

a. Give a code for the binary channel according to Huffman and determine the efficiency.

b. Give a code for the ternary channel according to Fano and determine the efficiency.

c. Which channel is given the preference (with codes according to a and b) one wants to keep the expected costs at a minimum, and how large are these costs then?

2.11. An information source generates zeros and ones with probabilities $p(0) = 0.8$ and $p(1) = 0.2$. The sequences of symbols are encoded with the code symbols 0, 1, 2, 3, 4 according to the table below:

message	code
1	0
01	1
001	2
0001	3
0000	4

a. Is this code uniquely decodable and instantaneously decodable?

b. Determine the average amount of information per code symbol.

c. What is the efficiency of this code?

2.12. A source generates messages from an alphabet consisting of eight symbols u_1 to u_8. The probabilities of occurring are 0.32, 0.24, 0.20, 0.09, 0.05, 0.04, 0.04 and 0.02.

a. Determine a suitable binary code with the help of the method of Shannon.

b. Determine a suitable three-valued code with the help of the method of Huffman.

c. Determine a suitable four-valued code with the help of the method of Fano.

d. Compare the efficiencies of the three codes.

e. Determine the amount of information per symbol for the binary code (round the probabilities off to one decimal place).

2.6 Solutions

2.1. *a.* Since successive symbols are statistically independent, we find for the amount of information per symbol

$$H(U) = -\sum_{i=1}^{3} p(u_i) \log p(u_i) = -\frac{7}{10} \log \frac{7}{10} - \frac{2}{10} \log \frac{2}{10} - \frac{1}{10} \log \frac{1}{10}$$

$$= -\frac{7}{10} \log 7 - \frac{2}{10} \log 10 = 1.15 \text{ bits/symbol.}$$

b. The probability $p(v_j)$ of a certain message v_j can, as a result of the fact that the generated symbols are statistically independent, be written as the product of the probabilities of the individual symbols, so that it follows for the probabilities of the $3^2 = 9$ possible messages of two symbols that

$$p(v_1) = p(u_1, u_1) = p(u_1) p(u_1) = 0.49,$$

$$p(v_2) = p(u_1, u_2) = p(u_1) p(u_2) = 0.14,$$

$$p(v_3) = p(u_1, u_3) = p(u_1) p(u_3) = 0.07,$$

$$p(v_4) = p(u_2, u_1) = 0.14,$$

$$p(v_5) = p(u_2, u_2) = 0.04,$$

$$p(v_6) = p(u_2, u_3) = 0.02,$$

$$p(v_7) = p(u_3, u_1) = 0.07,$$

$$p(v_8) = p(u_3, u_2) = 0.02,$$

$$p(v_9) = p(u_3, u_3) = 0.01.$$

c. With regard to a we have

$$H(V) = l H (U),$$

that is, a message of length l contains l times as much information as a message of length 1, when succeeding symbols are statistically independent. Thus

$$H(V) = 2 H(U) = 2.30 \text{ bits/message.}$$

With regard to b we have

$$H(V) = -\sum_{j=1}^{9} p(v_j) \log p(v_j)$$

$$= -0.49 \log 0.49 - 0.14 \log 0.14 - 0.07 \log 0.07$$

$$-0.14 \log 0.14 - 0.04 \log 0.04 - 0.02 \log 0.02$$

$$-0.07 \log 0.07 - 0.02 \log 0.02 - 0.01 \log 0.01$$

$$= 2.30 \text{ bits/message.}$$

Thus both outcomes are equal which agrees with theory.

d. The redundancy is equal to

$$\text{red} = 1 - \frac{H(U)}{\max\limits_{u} H(U)} = 1 - \frac{1.15}{\log 3} = 0.27.$$

e. For the information per second we have

$$H_t(U) = \frac{1}{t} H(U) \quad \text{bits/second,}$$

where t is the average duration of a symbol with

$$t = 0.7 \times 0.001 + 0.2 \times 0.002 + 0.1 \times 0.003 = 0.0014 \text{ sec.}$$

Thus

$$H_t(U) = \frac{1}{0.0014} 1.15 = 821.44 \text{ bits/second.}$$

2.2. *a.* The amount of information per symbol is

$$H(U) = -\sum_{i=1}^{8} p(u_i) \log p(u_i)$$

$$= -\frac{1}{2} \log \frac{1}{2} - \frac{1}{4} \log \frac{1}{4} - \frac{1}{8} \log \frac{1}{8} - \frac{1}{16} \log \frac{1}{16} - \frac{1}{32} \log \frac{1}{32}$$

$$- \frac{1}{64} \log \frac{1}{64} - 2 \times \frac{1}{128} \log \frac{1}{128} = 1\frac{63}{64} \text{ bit} = 1.98 \text{ bits.}$$

b. The probability of a zero can be determined as follows:

$$p(0) = \frac{\sum\limits_{i} p(u_i) \, c_{i0}}{\sum\limits_{i} p(u_i) \, l_i},$$

where c_{i0} indicates the number of zeros in code word u_i and l_i the number of symbols making up this code word. This gives

$$p(0) = \left\{ \frac{1}{2} 3 + \frac{1}{4} 2 + \frac{1}{8} 2 + \frac{1}{16} 1 + \frac{1}{32} 2 + \frac{1}{64} 1 + \frac{1}{128} 1 \right\} \frac{1}{3} = 0.8.$$

Then $p(1) = 1 - p(0) = 0.2$.

c. For the efficiency we have

$$\eta = \frac{H(U)}{L \log r},$$

where L is the average length of the code words. Because the coding is deterministic and one-to-one, no uncertainty is introduced, so that the amount of information of the source output symbols $H(U)$ is equal to the amount of information of the code words, thus also 1.98 bits. Substitution gives

$$\eta = \frac{1.98}{3 \times 1} = 0.66.$$

d. Fano and Shannon lead to the same code for the given probabilities. This becomes

symbol	probability	code ($r = 2$)
u_1	1/2 = 64/128	0
u_2	1/4 = 32/128	1 0
u_3	1/8 = 16/128	1 1 0
u_4	1/16 = 8/128	1 1 1 0
u_5	1/32 = 4/128	1 1 1 1 0
u_6	1/64 = 2/128	1 1 1 1 1 0
u_7	1/128 = 1/128	1 1 1 1 1 1 0
u_8	1/128 = 1/128	1 1 1 1 1 1 1

e. To determine the efficiency the average duration of the code words must first be calculated. This is

$$L = \frac{1}{2}\,1 + \frac{1}{4}\,2 + \frac{1}{8}\,3 + \frac{1}{16}\,4 + \frac{1}{32}\,5 + \frac{1}{64}\,6 + 2 \times \frac{1}{128}\,7 = 1\frac{63}{64} = 1.98.$$

The efficiency is then

$$\eta = \frac{1.98}{1.98 \times 1} = 1.$$

2.3. *a.* The application of the method of Fano for a ternary code proceeds by constantly dividing the symbols into three groups of approximately equal probabilities. In this manner the following code arises:

symbol	probability	code ($r = 3$)
u_1	$3/8 = 9/24$	0
u_2	$1/6 = 4/24$	1 0
u_3	$1/8 = 3/24$	1 1
u_4	$1/8 = 3/24$	2 0
u_5	$1/8 = 3/24$	2 1
u_6	$1/12 = 2/24$	2 2

b. The amount of information per symbol is

$$H(U) = -\frac{3}{8}\log\frac{3}{8} - \frac{1}{6}\log\frac{1}{6} - 3\times\frac{1}{8}\log\frac{1}{8} - \frac{1}{12}\log\frac{1}{12} = 2.39 \text{ bits.}$$

In addition, the average length of a code word must be determined. This is

$$L = \frac{3}{8}\,1 + \frac{1}{6}\,2 + 3\times\frac{1}{8}\,2 + \frac{1}{12}\,2 = 1.625.$$

In this case the efficiency is equal to

$$\eta = \frac{H(U)}{L \log r} = \frac{2.39}{1.625 \log 3} = 0.93.$$

2.4. *a.* The symbols have to be divided up into four groups of, as much as possible, equal probability. This gives code I:

symbol	probability	code I	code II
u_1	$1/4 = 15/60$	a	a
u_2	$1/5 = 12/60$	b	b
u_3	$1/6 = 10/60$	c a	c
u_4	$1/6 = 10/60$	c b	d a
u_5	$1/12 = 5/60$	d a	d b
u_6	$1/12 = 5/60$	d b	d c
u_7	$1/20 = 3/60$	d c	d d

Remark: Another code, namely code II, gives a better result (i.e. a code with a smaller average word length).

b. The amount of information is

$$H(U) = -\frac{1}{4}\log\frac{1}{4} - \frac{1}{5}\log\frac{1}{5} - 2\times\frac{1}{6}\log\frac{1}{6} - 2\times\frac{1}{12}\log\frac{1}{12} - \frac{1}{20}\log\frac{1}{20}$$

$$= 2.64 \text{ bits/symbol.}$$

The average length of the code words of code I is

$$L = (\tfrac{1}{4} + \tfrac{1}{5})\, 1 + (\tfrac{2}{6} + \tfrac{2}{12} + \tfrac{1}{20})\, 2 = \tfrac{31}{20} = 1.55.$$

The efficiency of this code is then

$$\eta = \frac{H(U)}{L \log r} = \frac{2.64}{1.55 \log 4} = 0.85.$$

With code II the efficiency is 0.95.

2.5. *a.* Using the method of Fano the best code is as follows:

symbol	probability	code ($r = 3$)
u_1	1/2 = 8/16	a
u_2	1/4 = 4/16	b
u_3	1/8 = 2/16	c a
u_4	1/16 = 1/16	c b
u_5	1/16 = 1/16	c c

b. The amount of information in this case is equal to

$$H(U) = -\tfrac{1}{2}\log\tfrac{1}{2} - \tfrac{1}{4}\log\tfrac{1}{4} - \tfrac{1}{8}\log\tfrac{1}{8} - 2\times\tfrac{1}{16}\log\tfrac{1}{16}$$

$$= \tfrac{15}{8} = 1.875 \text{ bit/symbol.}$$

The average length of the code words is

$$L = (\tfrac{1}{2} + \tfrac{1}{4})\, 1 + (\tfrac{1}{8} + 2\times\tfrac{1}{16})\, 2 = 1.25.$$

The efficiency of the code is therefore

$$\eta = \frac{1.875}{1.25 \log 3} = 0.95.$$

2.6. *a.* The method of Fano is chosen to find a suitable code:

symbol	probability	code ($r = 2$)
a	1/2 = 3/6	0
b	1/3 = 2/6	1 0
c	1/6 = 1/6	1 1

b. The amount of information is equal to

$$H(U) = -\tfrac{1}{2} \log \tfrac{1}{2} - \tfrac{1}{3} \log \tfrac{1}{3} - \tfrac{1}{6} \log \tfrac{1}{6} = 1.46 \text{ bits/symbol.}$$

The average length of the code words is

$$L = \tfrac{1}{2} \, 1 + (\tfrac{1}{3} + \tfrac{1}{6}) \, 2 = 1.5.$$

The efficiency of this code therefore comes out as

$$\eta = \frac{1.46}{1.5 \log 2} = 0.97.$$

c. The code can be improved by constantly taking two symbols together to form a new message symbol. Since these symbols are independent, the probability of a new symbol is simply the product of the probabilities of the original symbols. If the new symbols are arranged in order of decreasing probability the method of Fano gives the following code (more solutions are possible):

symbol	probability	code ($r = 2$)
aa	$1/4 = 9/36$	0 0
ab	$1/6 = 6/36$	0 1
ba	$1/6 = 6/36$	1 0 0
bb	$1/9 = 4/36$	1 0 1
ac	$1/12 = 3/36$	1 1 0 0
ca	$1/12 = 3/36$	1 1 0 1
bc	$1/18 = 2/36$	1 1 1 0
cb	$1/18 = 2/36$	1 1 1 1 0
cc	$1/36 = 1/36$	1 1 1 1 1

d. The amount of information of pairs of symbols is twice as large as that per symbol because the symbols are independent of each other, thus $H(V) = 2 \cdot 1.46 = 2.92$ bits.

The average length of the code words has now become

$$L = (1/4 + 1/6) \, 2 + (1/6 + 1/9) \, 3 + (1/12 + 1/12 + 1/18)4$$

$$+ \, (1/18 + 1/36) \, 5 = 2.97 \text{ per pair of symbols or } 1.49 \text{ per}$$
$$\text{original symbol}$$

by which $\eta = \dfrac{2.92}{2.97 \log 2} = 0.98.$

Taking symbols together thus leads to some improvement of the efficiency.

2.7. *a*. The method of Fano gives the following code

symbol	probability	code ($r = 3$)
u_1	$1/4 = 8/32$	a
u_2	$1/4 = 8/32$	b a
u_3	$1/8 = 4/32$	b b
u_4	$1/8 = 4/32$	c a
u_5	$1/16 = 2/32$	c b a
u_6	$1/16 = 2/32$	c b b
u_7	$1/16 = 2/32$	c c a
u_8	$1/32 = 1/32$	c c b
u_9	$1/32 = 1/32$	c c c

Another method is that of Huffman. This gives the following code:

symbol	probability				code ($r = 3$)
u_1	8/32	8/32	8/32	⇗16/32 (a)	b
u_2	8/32	8/32	8/32	8/32 (b)	c
u_3	4/32	4/32	⇗8/32 (a)⌐	8/32 (c)	a b
u_4	4/32	4/32	4/32 (b)		a c
u_5	2/32	⇗4/32 (a)⌐	4/32 (c)⌋		a a b
u_6	2/32	2/32 (b)			a a c
u_7	2/32 (a)⌐	2/32 (c)⌋			a a a a
u_8	1/32 (b)				a a a b
u_9	1/32 (c)⌋				a a a c

The amount of information per symbol is

$$H(U) = -2 \times \frac{1}{4} \log \frac{1}{4} - 2 \times \frac{1}{8} \log \frac{1}{8} - 3 \times \frac{1}{16} \log \frac{1}{16} - 2 \times \frac{1}{32} \log \frac{1}{32}$$

$$= 1 + \frac{3}{4} + \frac{3}{4} + \frac{5}{16} = 2.82 \text{ bits/symbol.}$$

The average length of the code for Fano is

$$L = 1 \times \frac{1}{4} + 2(\frac{1}{4} + \frac{1}{8} + \frac{1}{8}) + 3(\frac{1}{16} + \frac{1}{16} + \frac{1}{16} + \frac{1}{32} + \frac{1}{32}) = 2,$$

and for Huffman

$$L = 1\,(\frac{1}{4} + \frac{1}{4}) + 2(\frac{1}{8} + \frac{1}{8}) + 3(\frac{1}{16} + \frac{1}{16}) + 4\,(\frac{1}{16} + \frac{1}{32} + \frac{1}{32}) = \frac{15}{8} = 1.88.$$

The efficiency of the code according to Fano therefore comes out as

$$\eta_F = \frac{2.82}{2\log 3} = 0.89$$

and according to Huffman

$$\eta_H = \frac{2.82}{1.88\log 3} = 0.95.$$

It turns out that the method of Huffman leads to a more efficient code.

b. If the code symbol c may not be followed by c, it cannot be permitted that
– the combination cc appears in a code word,
– a code word ends with c, if one or more code words begin with c.
A possible code is the following one:

symbol	probability	code ($r = 3$)
u_1	1/4 $= 8/32$	a
u_2	1/4 $= 8/32$	b a
u_3	1/8 $= 4/32$	b b
u_4	1/8 $= 4/32$	c a a
u_4	1/16 $= 2/32$	c a b
u_6	1/16 $= 2/32$	c b a
u_7	1/16 $= 2/32$	c b b
u_8	1/32 $= 1/32$	c b c a
u_9	1/32 $= 1/32$	c b c b

The average length for this code is

$$L = \frac{1}{4} + 2\cdot(\frac{1}{4} + \frac{1}{8}) + 3\cdot(\frac{1}{8} + \frac{3}{16}) + 4\cdot(\frac{1}{32} + \frac{1}{32}) = 2.82.$$

The efficiency is therefore

$$\eta = \frac{2.82}{2.82\cdot\log 3} = 0.81.$$

2.8. *a,b.* Nine possibilities arise by combining the two answers. After rearranging in order of decreasing probability the following codes are found with the method of Fano:

combination	probability	code (a) ($r = 2$)	code (b) ($r = 3$)
yes, yes	0.30	0 0	a
no, yes	0.24	0 1	b a
yes, no	0.10	1 0 0	b b
yes, n.o.	0.10	1 0 1	c a
no, no	0.08	1 1 0 0	c b a
no, n.o.	0.08	1 1 0 1	c b b
n.o., yes	0.06	1 1 1 0	c c a
n.o., no	0.02	1 1 1 1 0	c c b
n.o., n.o.	0.02	1 1 1 1 1	c c c

c. The amount of information content of the message is the sum of the amounts of information with regard to question 1 and question 2. This leads to

$$H(V) = -0.5 \log 0.5 - 0.4 \log 0.4 - 0.1 \log 0.1$$

$$- 0.6 \log 0.6 - 0.2 \log 0.2 - 0.2 \log 0.2 = 2.73 \text{ bits.}$$

For the binary code in a the average length is

$$L = 2{\cdot}0.54 + 3{\cdot}0.20 + 4{\cdot}0.22 + 5{\cdot}0.04 = 2.76,$$

so that the efficiency is equal to

$$\eta_a = \frac{2.73}{2.76{\cdot}\log 2} = 0.99.$$

For the ternary code in b the average length is

$$L = 1{\cdot}0.30 + 2{\cdot}0.44 + 3{\cdot}0.26 = 1.96,$$

so that the efficiency here is equal to

$$\eta_b = \frac{2.73}{1.96{\cdot}\log 3} = 0.88.$$

On the basis of the efficiency the preference therefore goes to the binary code.

2.9. The symbols $u_1,\ldots,u_4$ are given with their respective probabilities in the table. A code according to the method of Gilbert and Moore is obtained by determining the length of the code word for each symbol and subsequently forming an increasing series α_i which is then developed into a binary series. For symbol u_1 we have

$$2^{-4} < p(u_1) = \tfrac{1}{10} < 2^{-3},$$

symbol	probability
u_1	0.1
u_2	0.3
u_3	0.2
u_4	0.4

so that a length of $t_1 = 5$ (binary) symbols is necessary. The value of $\alpha_1 = \tfrac{1}{2}\, p(u_1)$ becomes $\alpha_1 = 0.05$. If this is written out in binary notation and is broken off after 5 numbers then 00001 arises as the code word. The remaining three code words are determined in the same manner. The following table gives the result of the code:

	probability	l_i	α_i	code
u_1	0.1	5	0.05	0 0 0 0 1
u_2	0.3	3	0.25	0 1 0
u_3	0.2	4	0.50	1 0 0 0
u_4	0.4	3	0.80	1 1 0

To find the efficiency of this code the information content of the source must first be determined, which is given by

$$H(U) = - \sum_{i=1}^{4} p(u_i) \log p(u_i)$$

$$= -0.1 \log 0.1 - 0.3 \log 0.3 - 0.2 \log 0.2 - 0.4 \log 0.4$$

$$= 1.85 \text{ bits/symbol.}$$

The average length of the code words is

$$L = 0.1 \cdot 5 + 0.3 \cdot 3 + 0.2 \cdot 4 + 0.4 \cdot 3 = 3.4,$$

so that the efficiency becomes

$$\eta_G = \frac{1.85}{3.4 \cdot 1} = 0.54.$$

After rearranging the symbols according to increasing probability, a code according to Shannon arises in the following manner:

	probability	P_i	l_i	code ($r = 2$)
u_1	0.4	0	2	0 0
u_2	0.3	0.4	2	0 1
u_3	0.2	0.7	3	1 0 1
u_4	0.1	0.9	4	1 1 1 0

The average length of the code words is now

$$L_S = 0.4 \cdot 2 + 0.3 \cdot 2 + 0.2 \cdot 3 + 0.1 \cdot 4 = 2.4,$$

which yields an efficiency of

$$\eta_S = \frac{1.85}{2.4 \cdot 1} = 0.77.$$

Lastly, the code according to Huffman is found by constantly taking the two symbols with the least probability of occurring together. This results in the following code:

symbol	probability			code ($r = 2$)
u_4	0.4	0.4	0.6 (0)	1
u_2	0.3	0.3 (0)	0.4 (1)	0 0
u_3	0.2 (0)	0.3 (1)		0 1 0
u_1	0.1 (1)			0 1 1

The code according to Huffman has an average length

$$L_H = 0.4 \cdot 1 + 0.3 \cdot 2 + 0.2 \cdot 3 + 0.1 \cdot 3 = 1.9,$$

so that the efficiency yields

$$\eta_H = \frac{1.85}{1.9 \cdot 1} = 0.97.$$

It turns out that the method according to Huffman yields the best efficiency and that the Gilbert–Moore method is the worst of the three; this code does meet an extra requirement, however, namely that the ordering of the symbols remains the same.

2.10. *a.* It is necessary for both the Huffman and the Fano codes to arrange the symbols in order of decreasing probability of occurring. A binary Huffman code is the following for example:

symbol	probability						code ($r = 2$)
u_1	1/3	1/3	1/3	1/3	1/3	2/3 (0)	1
u_2	1/3	1/3	1/3	1/3	1/3 (0)	1/3 (1)	00
u_3	1/9	1/9	1/9	2/9 (0)	1/3 (1)		011
u_4	1/9	1/9	1/9 (0)	1/9 (1)			0100
u_5	1/27	2/27 (0)	1/9 (1)				01011
u_6	1/27 (0)	1/27 (1)					010100
u_7	1/27 (1)						010101

To determine the efficiency, the amount of information is first calculated

$$H(U) = -2\frac{1}{3}\log\frac{1}{3} - 2\frac{1}{9}\log\frac{1}{9} - 3\frac{1}{27}\log\frac{1}{27} = 2.29 \text{ bits/symbol.}$$

The average length is equal to

$$L_H = \frac{1}{3}(1 + 2) + \frac{1}{9}(3 + 4) + \frac{1}{27}(5 + 6 + 6) = 65/27 = 2.41.$$

The efficiency is therefore

$$\eta_H = \frac{2.29}{2.41 \cdot 1} = 0.95.$$

b. To determine a ternary Fano code the symbols are constantly divided up into three groups of approximately equal probabilities. The code symbols are denoted by a, b and c:

symbol	probability	code ($r = 3$)
u_1	1/3	a
u_2	1/3	b
u_3	1/9	c a
u_4	1/9	c b
u_5	1/27	c c a
u_6	1/27	c c b
u_7	1/27	c c c

The average length of this code is

$$L_F = \frac{1}{3}(1 + 1) + \frac{1}{9}(2 + 2) + \frac{1}{27}(3 + 3 + 3) = 13/9,$$

which leads to an efficiency of

$$\eta_F = \frac{2.29}{1.44 \cdot 1.58} = 1.00.$$

c. The ternary Fano code receives the preference on the basis of the efficiency. However, if one brings the cost per code symbol into account then one changes the criterion of maximum efficiency into a criterion of minimum expected costs. For the Huffman code we now have

$$\text{Costs}_H = (65/27) \cdot 1.80 = 4.33 \text{ (pounds).}$$

The cost of the Fano code is

$$\text{Costs}_F = (13/9) \cdot 2.70 = 3.90 \text{ (pounds).}$$

The Fano code therefore also enjoys the preference in that case.

2.11. *a.* On the side of the decoder one receives a series of numbers, which one can convert directly into a sequence of binary symbols without having to wait for the next symbol. The code is thus directly decodable. The code is also instantaneously decodable since an unambiguous one-to-one relation exists between the binary source symbols and the code symbols even if they occur in a sequence of symbols.

b. The information content per code symbol can be determined if the probabilities of the code symbols have been calculated. With the given probabilities of the binary symbols these can be determined as follows, for example

$$p(3) = p(0)^3 \, p(1) = 0.8^3 \, 0.2 = 0.1024.$$

One thus finds the following probabilities:

code symbols	probability
0	0.2
1	0.16
2	0.128
3	0.1024
4	0.4096

The information content per code symbol is then

$$H(U) = -0.2 \log 0.2 - 0.16 \log 0.16 - 0.128 \log 0.128$$

$$- 0.1024 \log 0.1024 - 0.4096 \log 0.4096 = 2.13 \text{ bits/symbol.}$$

c. The efficiency is as follows:

$$\eta = \frac{H(U)}{L \log r} = \frac{2.13}{1 \cdot \log 5} = 0.92.$$

2.12. *a.* Shannon's method uses the cumulative probability P_k. The length of the code words follows from

$$\log \frac{1}{p_k} \leq l_k \leq \log \frac{1}{p_k} + 1.$$

The final result is as follows:

symbol	p_i	P_i	l_i	code ($r = 2$)
u_1	0.32	0	2	0 0
u_2	0.24	0.32	3	0 1 1
u_3	0.20	0.56	3	1 0 0
u_4	0.09	0.76	4	1 1 0 0
u_5	0.05	0.85	5	1 1 0 1 1
u_6	0.04	0.90	5	1 1 1 0 1
u_7	0.04	0.94	5	1 1 1 1 0
u_8	0.02	0.98	6	1 1 1 1 1 1

b. For three code symbols the number of message symbols must be equal to $3 + 2k$ to come to an optimal Huffman code. We must therefore add a fictitious symbol that has a probability of zero. We then find

symbol	probability				code ($r = 3$)
u_1	0.32	0.32	0.32	→0.44 (a)	b
u_2	0.24	0.24	0.24	0.32 (b)	c
u_3	0.20	0.20	0.20 (a)	0.24 (c)	a a
u_4	0.09	0.09	→0.15 (b)		a c
u_5	0.05	→0.06 (a)	0.09 (c)		a b b
u_6	0.04	0.05 (b)			a b c
u_7	0.04 (a)	0.04 (c)			a b a a
u_8	0.02 (b)				a b a b
u_9	0.00 (c)				– – – –

c. Fano's method for a four-valued code is based on the process of splitting the symbols into four groups of approximately equal probabilities per group

symbol	prob.	code ($r = 4$)
u_1	0.32	a
u_2	0.24	b
u_3	0.20	c
u_4	0.09	d a
u_5	0.05	d b
u_6	0.04	d c
u_7	0.04	d d a
u_8	0.02	d d b

d. It is not necessary to determine the whole efficiency for a comparison of the three codes. One can make do with a comparison of $L \log r$. One finds

code	L	$\log r$	$L \log r$
I	3.09	1	3.09
II	1.65	1.58	2.61
III	1.30	2	2.60

One can conclude on the basis of this that the Fano code is the best, albeit with a marginal difference with respect to the Huffman code.

e. With the binary code found in a it follows for the probability of a zero that

$$p(0) = \frac{1}{L} \sum_{i=1}^{8} p(u_i)\, c_i(0),$$

where $c_i(0)$ is the number of zeros in the code word for symbol u_i. It now follows that

$$p(0) = \frac{1}{3.09} \left[0.32 \times 2 + 0.24 \times 1 + 0.20 \times 2 + 0.09 \times 2 + 0.05 \times 1 \right.$$

$$\left. + 0.04 \times 1 + 0.04 \times 1 + 0.02 \times 0 \right] = 0.5,$$

and thus $p(1) = 1 - p(0) = 0.5$.

The information content per symbol of the binary code is therefore

$$H(U) = -0.5 \log 0.5 - 0.5 \log 0.5 = 1 \text{ bit.}$$

3

The discrete information source with memory

3.1 Markov processes

In Chapter 2 it was assumed that the information sources have no memory, which means that successive symbols in a message generated by the source are statistically independent. In many practical applications this is not the case and the probability of a symbol occurring in a message will depend on a finite number of preceding symbols. In this case one speaks of an information source with memory. A sequence generated by such a source can be regarded as a so-called *Markov chain*. Before discussing the discrete information source with memory, some of the properties of Markov chains will be considered in more detail.

We will consider a sequence of discrete stochastic variables $\mathbf{u}_1, \mathbf{u}_2, \ldots, \mathbf{u}_{n-1}$. These can be symbols from a discrete information source for example, or the quantified samples of a stochastic signal. The first $\mathbf{u}_1$ is related to all the symbols which the first symbol can be, $\mathbf{u}_2$ to all symbols which the second symbol can be, etc. In the case of memoryless sources all symbols of the generated messages can be all symbols of the source alphabet. In the case of sources with memory it can happen that the second symbol can only be an element of a limited subset of the source alphabet, due to the occurrence of the first symbol, etc. We are now interested in the probability distribution of $\mathbf{u}_n$ which, for a *Markov chain of order k*, is dependent on the values of the k preceding stochastic variables $\mathbf{u}_{n-k}, \mathbf{u}_{n-(k-1)}, \ldots, \mathbf{u}_{n-1}$, where k is chosen in such a way that:

a. the values of the stochastic variables that precede $\mathbf{u}_{n-k}$ have no influence on the probability distribution of $\mathbf{u}_n$,

b. k is the minimum value for which a is valid.

The conditional probability of a value u_n of the stochastic variable $\mathbf{u}_n$, given all preceding values $u_{n-1}, u_{n-2}, \ldots$, is then equal to $P(u_n/u_{n-k}, u_{n-k+1}, \ldots, u_{n-1})$. Because the combination of values $S_i = (u_{n-k}, u_{n-k+1}, \ldots, u_{n-1})$ forms the conditional part of the conditional probability distribution of $\mathbf{u}_n$, S_i is called the *state of the Markov chain*. If the value u_n occurs, then the Markov chain goes over into a new state S_j:

$$S_j = (u_{n-(k-1)}, u_{n-(k-2)}, \ldots, u_n).$$

One then speaks of the *transition* of state S_i to state S_j. Besides giving the conditional probability distribution of $\mathbf{u}_n$, one can also characterise the Markov chain by giving the matrix of the transition probabilities for the various states. We will denote the probability that a Markov chain goes over from state S_i into state S_j by $P(j/i)$.

In Section 2.1 attention was paid to the case where l symbols generated by an information source were constantly regarded as a new message. We are then actually dealing with a new information source, which has an alphabet V, the symbols of which are composed from the symbols of the original alphabet U. If the combination S_i of k values of the stochastic variables is regarded in an analogous manner as the value S_i of one new stochastic variable, then the probability of state S_j is dependent only on the state S_i. Through this, the Markov chain of the old order k is reduced in terms of the symbols to a Markov chain of order 1 in terms of the states. Although the Russian mathematician Markov used chains where the order k could be greater than one, in the current mathematical literature Markov chains of the order $k = 1$ are used exclusively. In information theory, however, it is precisely the higher-order Markov chains that are of importance, because the description of the information source then sometimes becomes simpler.

If we assume here that every stochastic variable $\mathbf{u}_i$ has m possible outcomes, a Markov chain of order k can find itself in m^k different states as every state is determined by a sequence of k symbols, each chosen from m possibilities. Since after each state a choice can be made from m possibilities, there are m^{k+1} conceivable transitions with an equal number of transition probabilities. From every group of m transition probabilities, there will be $m - 1$ probabilities that can be chosen freely. The remaining probability is then fixed, since the sum of the transition probabilities is equal to one. The m^{k+1} transition probabilities are therefore determined by $m^{k+1} - m^k$ freely chosen transition probabilities.

One can depict the so defined states with their transition probabilities in a *state diagram*. One can consider a sequence of statistically independent symbols as the most simple case. As k is then equal to zero, one could speak of a Markov chain of order 0. This is actually a degeneration of a Markov chain, which is by definition based on the existence of transition probabilities. Such a chain will have just one state *S*; after each "transition" the chain returns to the same state. The number of possible transitions is equal to the number of symbols that can be chosen. For 3 symbols (a, b, c) this chain can be represented as shown in Figure 3.1.

For a Markov chain of order 1, the number of states is equal to the number of symbols. If this number is 3, namely a, b and c, then the number of transitions is $3^2 = 9$, namely a → a, a → b, a → c, b → a, etc. Such a Markov chain can be represented as in Figure 3.2.

The states S_1, S_2 and S_3 can be denoted by a, b and c in this example. Naturally, for every state S_i, i = 1, 2, and 3, we have that

$$P(S_1/S_i) + P(S_2/S_i) + P(S_3/S_i) = 1. \tag{3.1}$$

The probabilities of the 3 states S_1, S_2 and S_3 can be determined from the transition probabilities:

Figure 3.1. State diagram for a Markov chain of order 0.

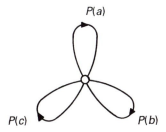

Figure 3.2. State diagram for a Markov chain of order 1.

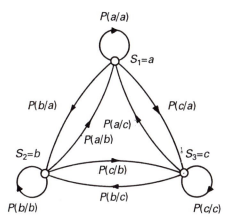

$$P(S_i) = P(S_1) \cdot P(S_i/S_1) + P(S_2) \cdot P(S_i/S_2) + P(S_3) \cdot P(S_i/S_3), \qquad (3.2)$$

for $i = 1,2,3$.

The following should be noted regarding this expression. In Section 1.2 (see equation (1.14)) Bayes' theorem, which indicated how the conditional probabilities $p(x_i/y_j)$ can be calculated on the basis of given conditional probabilities $q(y_j/x_i)$, was treated. This theorem may not be used for the transition probabilities in Markov chains however. The reason is that Bayes' theorem, in reality, assumes that for all i and j

$$p(x_i, y_j) = p(y_j, x_i).$$

However, this is generally not the case for Markov chains. We have instead that

$$P(S_i, S_j) \neq P(S_j, S_i). \qquad (3.3)$$

The reason for this is that the time aspect plays an important role for Markov chains and therefore also the ordering of the symbols. If we think of the written language (which can in fact be regarded as a Markov process where the appearance of letters is determined by the preceding letters) then it is clear that the probability of the letter pair (q,u) will not be equal to the probability of the letter pair (u,q).

The importance of the Markov chains is due to the fact that k, the number of stochastic variables the values of which determine the probability of transition to the next stochastic variable $\mathbf{u}_n$, is finite, so that the past does not have to be taken into account *ad infinitum*.

Example 3.1

Consider an information source that generates a Markov chain. The source alphabet is $U = \{0,1\}$. The following transition probabilities are further assumed:

$$P(0/00) = P(1/11) = 0.8,$$

$$P(1/00) = P(0/11) = 0.2,$$

$$P(0/01) = P(0/10) = P(1/01) = P(1/10) = 0.5.$$

It can be concluded that we are dealing with a Markov process of order 2 here. The occurrence of a symbol is determined by the occurrence of the 2 preceding symbols. There are thus four states: 00, 01, 10, 11. The state diagram is depicted in Figure 3.3. It can be concluded from the figure, that we cannot directly reach every state from every other state. Thus we can get to S_2 from S_1, but not to S_3 or S_4. Sometimes we can go from one state to another but not back: we can go to S_2 from S_1, but not from S_1 directly to S_2. The marginal probabilities of the states S_1 to S_4 can be calculated on the basis of equation (3.2).

$$P(S_1) = P(S_1){\cdot}P(S_1/S_1) + P(S_2){\cdot}P(S_1/S_2) + P(S_3){\cdot}P(S_1/S_3) + P(S_4){\cdot}P(S_1/S_4)$$

$$= P(S_1){\cdot}0.8 \quad + P(S_2){\cdot}0 \quad + P(S_3){\cdot}0.5 \quad + P(S_4){\cdot}0$$

$$= 0.8\ P(S_1) + 0.5\ P(S_3).$$

In the same manner it can be derived that

$$P(S_2) = P(S_1){\cdot}0.2 + P(S_2){\cdot}0 + P(S_3){\cdot}0.5 + P(S_4){\cdot}0$$
$$= 0.2\ P(S_1) + 0.5\ P(S_3),$$
$$P(S_3) = P(S_1){\cdot}0 + P(S_2){\cdot}0.5 + P(S_3){\cdot}0 + P(S_4){\cdot}0.2$$
$$= 0.5\ P(S_2) + 0.2\ P(S_4),$$
$$P(S_4) = P(S_1){\cdot}0 + P(S_2){\cdot}0.5 + P(S_3){\cdot}0 + P(S_4){\cdot}0.8$$

Figure 3.3. State diagram for Example 3.1.

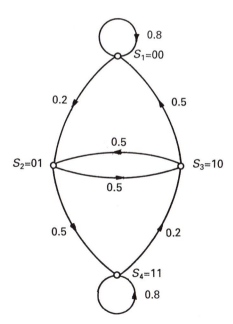

$$= 0.5 \, P(S_2) + 0.8 \, P(S_4).$$

Solving these four relations with four unknowns one finds

$$P(S_1) = P(S_4) = 5/14,$$

$$P(S_2) = P(S_3) = 2/14. \hspace{4em} \triangle$$

For reasons of completeness it is mentioned here that sometimes state diagrams are replaced by *trellis diagrams*. In fact a trellis diagram is a state diagram augmented by a time axis so that the state changes can be seen as a function of time. Compare Figure 3.4(a) and (b). Each Markov chain then corresponds to a specific path in the trellis diagram.

A few of the properties of Markov chains will be mentioned in this chapter. Two properties that find applications in information theory are mentioned here without derivation:

Figure 3.4 (a) State diagram; (b) trellis diagram.

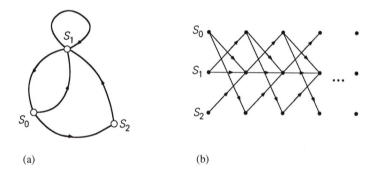

(a)   (b)

Figure 3.5. Non-ergodic Markov chain.

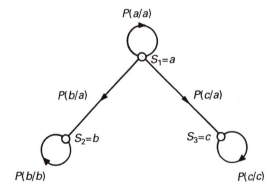

a. A part of a Markov chain is also a Markov chain.

b. A Markov chain that is passed through in the reverse direction is also a Markov chain.

A few limitations will be imposed on the Markov chains being considered here. In the first place we require that at every transition the matrix of transition probabilities is the same. One then calls the transition probabilities *stationary* and the Markov chain *homogeneous*. We further require that the Markov chain itself is also *stationary*, that is, that the probabilities of the states that the Markov chain can find itself in do not change. Lastly, we will further limit ourselves to *ergodic* Markov chains; that means no matter what state it finds itself in, from each state one can eventually reach any other state.

3.2 The information of a discrete source with memory

There is a certain amount of dependence between the successive symbols in the case of discrete sources with memory. The dependence here mentioned can extend over arbitrarily long sequences of symbols. However, one can often assume that this dependence extends over a limited number of symbols, making it possible to use the Markov chain introduced in Section 3.1 as a model for the information source with memory. The order of the Markov chain can be determined experimentally. Estimation techniques exist to do this, but as this is beyond the scope of this book no further mention will be made of this here. We assume, henceforth, that the information source is ergodic. This makes it possible to say, at a certain point in time, something about the probability of a certain symbol. This probability is then equal to the probability with which this symbol appears in an (infinitely) long sequence of successive symbols. A discrete information source with memory serves as a model for the written language for example, and thus for every stochastic sequence that is derived therefrom. Another example is a sampled quantified signal, where the sampling frequency is not chosen correctly and where independence therefore arises between the (quantified) samples.

I Amount of information for a first-order Markov chain

For a first-order Markov chain we have that the number of symbols u_i, $i = 1,2,...,m$, is equal to the number of states S_i, as we saw in the previous section. We will now consider transitions from an arbitrary symbol u_{1i} at a time t_1, to a symbol u_{2j} at a time t_2, where $i,j = 1,2,...,m$. The conditional probability, the probability of transition from u_{1i} to u_{2j}, can be denoted by

$p(u_{2j}/u_{1i})$. The amount of information belonging to an arbitrary transition is now given by (compare with Definition 1.3)

$$H(U_2/U_1) = -\sum_{i=1}^{m} \sum_{j=1}^{m} P(u_{1i}, u_{2j}) \log P(u_{2j}/u_{1i}). \tag{3.4}$$

For the joint amount of information of two symbols we have (compare with Definition 1.2)

$$H(U_1, U_2) = -\sum_{i=1}^{m} \sum_{j=1}^{m} P(u_{1i}, u_{2j}) \log P(u_{1i}, u_{2j}). \tag{3.5}$$

We further have that

$$H(U_1, U_2) = H(U_1) + H(U_2/U_1). \tag{3.6}$$

The amount of information in a message of length two is therefore equal to the sum of the amount of information of the first symbol and the conditional amount of information of the second symbol given the first symbol. As was derived in Theorem 1.2, we have

$$H(U_2/U_1) \leq H(U_2), \tag{3.7}$$

and it therefore follows from equation (3.6) that

$$H(U_1, U_2) \leq H(U_1) + H(U_2). \tag{3.8}$$

Equality holds if successive symbols are statistically independent, thus if the source is memoryless. Since the source is stationary and ergodic, $H(U_1) = H(U_2) = H(U)$ so that we can write

$$H(U_1, U_2) \leq 2H(U). \tag{3.9}$$

The amount of information in a message consisting of two symbols is therefore smaller for a source with memory than for a source without memory.

II Amount of information for higher-order Markov chains
Because an arbitrary discrete information source can generate a Markov chain of order $k > 1$, it is desirable to extend the foregoing to sources with an arbitrarily large memory. We will consider the conditional amount of information $F_N(U)$ of a symbol u_N, in the case where $N - 1$ preceding symbols are known, that is:

$$F_N(U) = H(U_N/U_{N-1}, \ldots, U_2, U_1). \tag{3.10}$$

This amount of conditional information has a number of properties. The first property is

$$H(U_N/U_{N-1},\ldots,U_2,U_1) \leq H(U_N/U_{N-1},\ldots,U_2). \tag{3.11}$$

What this property actually says is that the knowledge that is delivered by the first symbol cannot lead to an increase in the uncertainty about the Nth symbol, but decreases it or leaves it unchanged.

Theorem 3.1
The conditional amount of information $F_N(U) = H(U_N/U_{N-1},\ldots,U_1)$ of the Nth symbol in the case where the preceding $N-1$ symbols are known is a monotonic decreasing function of N. That is,

$$H(U_N/U_{N-1},\ldots,U_1) \leq H(U_{N-1}/U_{N-2},\ldots,U_1) \leq \ldots$$

$$\ldots \leq H(U_2/U_1) \leq H(U_1). \tag{3.12}$$

Proof
Since the source is stationary, the conditional amounts of information are independent of the position of the Nth symbol in the chain. Thus for example

$$H(U_{N-1}/U_{N-2},\ldots,U_1) = H(U_N/U_{N-1},\ldots,U_2),$$

and on the basis of the first property (see equation (3.11)) it follows directly that

$$H(U_N/U_{N-1},\ldots,U_1) \leq H(U_{N-1}/U_{N-2},\ldots,U_1),$$

so that

$$F_N(U) \leq F_{N-1}(U) \leq \ldots \leq F_2(U) \leq F_1(U). \qquad \square$$

This theorem thus says that for an increasing value of N, the conditional amount of information becomes smaller or at most remains the same. Because every amount of information is always greater than or equal to 0, it follows that $F_N(U)$ approaches a bounding value, which we denote by

$$H_\infty(U) = \lim_{N\to\infty} F_N(U) = \lim_{N\to\infty} H(U_N/U_{N-1},\ldots,U_1) \text{ bits/symbol.} \tag{3.13}$$

Obviously, if the source alphabet U consists of m symbols,

$$0 \leq H_\infty(U) \leq \log m. \tag{3.14}$$

The amount $H_\infty(u)$ is now defined as the amount of information of a discrete information source with memory. The memory may therefore be of unlimited length. If the source generates a Markov chain of order k then this means that

$$P(u_N/u_{N-1},...,u_1) = P(u_N/u_{N-k},...,u_{N-1}). \tag{3.15}$$

For the conditional amount of information we then have

$$H(U_N/U_{N-1},...,U_1) \quad = H(U_N/U_{N-k},...,U_{N-1})$$

$$= H(U_{k+1}/U_k,...,U_1). \tag{3.16}$$

Thus for increasing N there is now no increase in the memory length, because it remains limited to k. This means that from $N = k + 1$ the quantity $F_N(U)$ remains equal to $F_{k+1}(U)$ and thus does not decrease any more. It then follows that the limit value $H_\infty(U)$ for a Markov chain of order k is equal to

$$H_\infty(U) = F_{k+1}(U) = H(U_{k+1}/U_k,...,U_2,U_1). \tag{3.17}$$

If the source is memoryless then $k = 0$, so that $H_\infty(U)$ is equal to $H(U)$. For increasing order k, $H_\infty(U)$ will become continually smaller.

Besides individual symbols, one will often also consider messages that are made up of N symbols for example. One can derive the amount of information per symbol on the basis of the amount of information per message $H(V)$. The quantity $H(V)$ is defined as

$$H(V) = H(U_1,U_2,...,U_N) \quad \text{bits/message.} \tag{3.18}$$

The amount of information per symbol is now defined as

$$H_N(U) = \frac{1}{N}H(V) = \frac{1}{N}H(U_1,U_2,...,U_N) \quad \text{bits/symbol.} \tag{3.19}$$

If the symbols u_i are statistically independent, then

$$H_N(U) = \frac{1}{N}\sum_{i=1}^{N} H(U_i) = \frac{1}{N}N\,H(U) = H(U).$$

If the symbols are dependent, then

$$H_N(U) = \frac{1}{N}\left[(H(U_1) + H(U_2/U_1) + ... + H(U_N/U_{N-1},...,U_2,U_1)\right]$$

$$= \frac{1}{N} \sum_{j=1}^{N} F_j(U). \tag{3.20}$$

As was the case for $F_N(U)$, $H_N(U)$ is also monotonic decreasing and reaches the bounding value $H_\infty(U)$ for increasing N.

Theorem 3.2
If $H(V)$ is the amount of information for a message of length N then the amount of information per symbol, defined by $H_N(U) = H(V)/N$, is monotonic decreasing. We further have that

$$\lim_{N \to \infty} H_N(U) = H_\infty(u). \tag{3.21}$$

Proof
Making use of equations (3.12) and (3.20) it follows for $H(V)$ that

$$H(V) = N\,H_N(U) = H(U_1) + H(U_2/U_1) + \ldots + H(U_N/U_{N-1},\ldots,U_2,U_1)$$

$$\geq N\,H(U_N/U_{N-1},\ldots,U_2,U_1),$$

or with formula (3.10)

$$H_N(U) \geq F_N(U). \tag{3.22}$$

We can now write

$$H(V) = H(U_1,\ldots,U_N) = H(U_1,\ldots,U_{N-1}) + H(U_N/U_{N-1},\ldots,U_1)$$

or

$$N\,H_N(U) = (N-1)H_{N-1}(U) + F_N(U)$$

$$\leq (N-1)H_{N-1}(U) + H_N(U).$$

Therefore

$$(N-1)\,H_N(U) \leq (N-1) \cdot H_{N-1}(U),$$

or $\qquad H_N(U) \leq H_{N-1}(U), \tag{3.23}$

which proves that $H_N(U)$ is monotonic decreasing. Since $H_N(U) \geq 0$, it must converge to a limit . This limit is also $H_\infty(U)$. As was previously shown,

$$H_N(U) = \frac{1}{N} \sum_{j=1}^{N} F_j(U).$$

Since $F_j(U)$ converges to $H_\infty(U)$ for $j \to \infty$ it follows that

$$\lim_{N \to \infty} H_N(U) = \lim_{N \to \infty} \frac{1}{N} \sum_{j=1}^{N} F_j(U) = \frac{1}{N} [N\, H_\infty(U)] = H_\infty(U),$$

which is in agreement with equation (3.21).   □

As follows from equation (3.13) and Theorem 3.2, both $F_N(U)$ and $H_N(U)$ converge to the same limit. As has been shown, $H_N(U) \geq F_N(U)$, so that $H_N(U)$ is thus a worse approximation to the actual amount of information $H_\infty(U)$. An advantage of $H_N(U)$, however, is its simplicity.

Example 3.2
Assume that the various values of $F_j(U)$ are known for 26 different symbols (e.g. related to language):

$F_1 = 4.15$, then	$H_1 = F_1$	$= 4.15$
$F_2 = 2.99$	$H_2 = \frac{1}{2}(F_1 + F_2)$	$= 3.75$
$F_3 = 2.56$	$H_3 = \frac{1}{3}(F_1 + F_2 + F_3)$	$= 3.23$
$F_4 = 2.20$	$H_4 = \frac{1}{4}(F_1 + ...)$	$= 2.98$
$F_5 = 1.95$	$H_5 = {-}{-}{-}$	$= 2.77$
$F_6 = 1.72$	$H_6 = {-}{-}{-}$	$= 2.60$
$F_7 = 1.63$	$H_7 = {-}{-}{-}$	$= 2.46$
$F_8 = 1.60$	$H_8 = {-}{-}{-}$	$= 2.35$
$-----$	$---$	$----$

From Figure 3.6 it follows that the limit of $F_N(U)$ and $H_N(U)$ is approximately

$$H_\infty(U) = 1.50 \ \text{bits/symbol},$$

whereas $\max_u H(U) = \log 26 = 4.70$ bits/symbol.   △

3.3 Coding aspects

As is the case for the source without memory, one can also determine how large the number of most probable messages is that a source with memory can generate. One would expect this number to be smaller than with a source without memory due to the memory. This indeed proves to be the case. One can prove that if ℓ symbols are taken together instead N, then, for increasing ℓ, $-\log \rho(v)/\ell$ tends to $H_\infty(U)$.

Theorem 3.3
Given any $\varepsilon > 0$ and $\delta > 0$, we can find an ℓ_0 such that the sequences of any length $\ell \geq \ell_0$ fall into two classes:
(i)   A set S' whose total probability is less than ε.
(ii) The remainder set S, all of whose members have probabilities satisfying the inequality

$$\left| \frac{-\log \rho(v)}{\ell} - H_\infty(U) \right| < \delta. \tag{3.24}$$

This group is the set of *most probable messages*.   □

This theorem corresponds with the Shannon-McMillan theorem (see Theorem 2.3) which is related to the memoryless sources. Clearly, in that case also equation (3.24) becomes identical to equation (2.24) since then $H_\infty(U) = H(U)$.
The group S of most probable messages exists with probability $p(S) > 1 - \varepsilon$, where each message has a probability of

Figure 3.6. $F_N(U)$ and $H_N(U)$ as function of N.

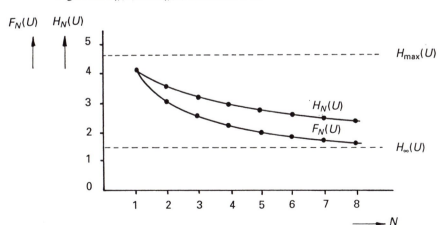

$$p(v) \approx 2^{-l \cdot H_\infty(U)}. \tag{3.25}$$

The number of most probable messages is approximately

$$M_\infty = \frac{1}{p(v)} \approx 2^{l \cdot H_\infty(U)}. \tag{3.26}$$

Since $H_\infty(U) \leq H(U)$, the number of most probable messages M_∞ of a source with memory is smaller than or at most equal to the number of probable messages of a source without memory.

In Section 2.1 the *redundancy* is defined as

$$\text{red} = 1 - \frac{H(U)}{\max\limits_{u} H(U)} = 1 - \frac{H(U)}{\log n}. \tag{3.27}$$

This measure thus gives an impression of the quality of a source without memory. In this chapter, it turned out that dependence between symbols also leads to a smaller information content. We can denote this loss of information by the *dependence redundancy*:

$$\text{red}_\infty = 1 - \frac{H_\infty(U)}{H(U)}, \tag{3.28}$$

where $H_\infty(U)$ is the amount of information of a source with memory and $H(U)$ the amount of information of a source without memory, the symbols of which have the same probabilities as those of the source with memory. Finally, we will give the definition of *total redundancy*:

$$\text{red}_{\text{tot}} = 1 - \frac{H_\infty(U)}{\max\limits_{u} H(U)} = 1 - \frac{H_\infty(U)}{\log n}. \tag{3.29}$$

Example 3.3
In Example 3.2 with max $H(U) = 4.70$ bits/symbol, $H(U) = 4.15$ bits/symbol and $H_\infty(U) = 1.50$ bits/symbol, we find the following for the various redundancy measures:

$$\text{red} \quad = 1 - \frac{4.15}{4.70} = 0.12,$$

$$\text{red}_\infty \quad = 1 - \frac{1.50}{4.15} = 0.64,$$

and $\quad \text{red}_{\text{tot}} = 1 - \frac{1.50}{4.70} = 0.68. \qquad \triangle$

With respect to sources with memory there are thus three different measures of redundancy, and each expresses some specific property of the underlying dependence between the symbols. For a Markov source of order 0 we find red = red$_{tot}$ and red$_\infty$ = 0.

One can obtain a source coding theorem for an information source with memory that corresponds with Shannon's first coding theorem derived in Section 2.4 (Theorem 2.5) by setting $H_\infty(U)$ instead of $H(U)$. Since the proof goes along similar lines to that of Theorem 2.5 it will not be given here.

Theorem 3.4
For a discrete information source with memory with an amount of information $H_\infty(U)$, where messages of length ℓ are encoded in code words of length L from a code alphabet with size r, it is the case that the probability P_e that there is a message for which there is no code word can be made arbitrarily small ($P_e < \varepsilon$) if L satisfies

$$L \log r \geq \ell H_\infty(U), \tag{3.30}$$

and ℓ is sufficient large. □

As follows from Example 3.3 among others, the dependence redundancy can be large enough that it becomes desirable to remove it through coding. It has been indicated for a memoryless source how one develops coding methods to obtain a suitable code which minimises the average length of the code words and thus minimises the redundancy in the code symbols at the same time. One way to remove the dependence redundancy for a source with memory is to apply these coding methods on messages of length *l* instead of on individual symbols. This approach is thus based on alphabet extension. One can choose the length *l* together with the order of the Markov chain.

Example 3.4
Consider an information source that generates a first-order Markov chain. The source alphabet is $U = \{A,B,C\}$. The following transition probabilities are given:

$$p(A/A) = \tfrac{1}{2}, \qquad p(B/A) = \tfrac{1}{2}, \qquad p(C/A) = 0,$$
$$p(A/B) = \tfrac{1}{4}, \qquad p(B/B) = 0, \qquad p(C/B) = \tfrac{3}{4},$$
$$p(A/C) = \tfrac{1}{3}, \qquad p(B/C) = \tfrac{1}{3}, \qquad p(C/C) = \tfrac{1}{3}.$$

The marginal probabilities follow from the following equations.

$$
\begin{cases}
p(A) = & \tfrac{1}{2}p(A) + \tfrac{1}{4}p(B) + \tfrac{1}{3}p(C), \\[2mm]
p(B) = & \tfrac{1}{2}p(A) + \qquad\quad \tfrac{1}{3}p(C), \\[2mm]
p(C) = & \qquad\quad \tfrac{3}{4}p(B) + \tfrac{1}{3}p(C), \\[2mm]
p(A) + p(B) + p(C) = 1.
\end{cases}
$$

The results are $p(A) = \tfrac{10}{27}$, $p(B) = \tfrac{8}{27}$ and $p(C) = \tfrac{9}{27}$.

Assume that just two code symbols are combined, on the basis of the transition probabilities the corresponding joint probabilities can be found. These joint probabilities together with the code words obtained from application of the Fano method for coding, in this case with $r = 2$, are given in the next table.

The average code word length is $L = \tfrac{75}{27} \approx 2.78$ or 1.39 per symbol. Computing the joint amount of information yields $H(U_1,U_2) = 2.72$. Thus the efficiency is $\eta = H(U_1,U_2)/L = 2.72/2.78 \approx 0.98$.

If Fano were applied to the separate source symbols the average code word length would be $L = \tfrac{44}{27} \approx 1.63$, which is larger than the 1.39 per symbol given above. The efficiency, as can be shown, has been decreased to $\eta = 0.97$.

	probability	code word
BC	$\tfrac{6}{27}$	0<u>0</u>
AA	$\tfrac{5}{27}$	0<u>1</u>
AB	$\tfrac{5}{27}$	10<u>0</u>
CA	$\tfrac{3}{27}$	1<u>01</u>
CB	$\tfrac{3}{27}$	11<u>0</u>
CC	$\tfrac{3}{27}$	111<u>0</u>
BA	$\tfrac{2}{27}$	1111
AC	0	—
BB	0	—

$\triangle$

One can also try to model the information source in another manner and bring it back to a source without memory. An example of this is the so-called run length code, which is applied among others for the coding of documents. With this method a document is scanned line by line and digitised into white or black image points. In this manner one obtains a binary information source with memory which generates two symbols, 0 and 1, corresponding with white and black. By approximation one can assume that the dependence between the symbols can be characterised by a Markov chain of the first order, where the probability $p(0)$ of a white image point is considerably larger than the probability $p(1)$ of a black image point. We now consider sequences of successive zeros and ones respectively, and determine the length of these sequences or runs. Each sequence of k zeros ending with a one can be regarded as a sequence of $k - 1$ transitions of white to white, followed by one transition of white to black. The probability of such a sequence is then

$$P_k(0) = P(0/0)^{k-1} \cdot P(1/0).$$ (3.31)

Similarly, the probability of a sequence of k ones (black pixels) is

$$P_k(1) = P(1/1)^{k-1} \cdot P(0/1).$$ (3.32)

This can be used to determine the average length of a sequence of white and black pixels respectively. We then find

$$\overline{k(0)} = \sum_{k=1}^{\infty} k \, P(0/0)^{k-1} \cdot P(1/0) = \frac{1}{P(1/0)},$$ (3.33)

$$\overline{k(1)} = \sum_{k=1}^{\infty} k \, P(1/1)^{k-1} \cdot P(0/1) = \frac{1}{P(0/1)}.$$ (3.34)

The original information source is now regarded as an information source that generates whole numbers, namely the run lengths, with probabilities $P_k(0)$ and $P_k(1)$ for $k = 1,2,\ldots,\infty$. In practice, one will tolerate a certain maximum length K and regard longer sequences as multiples of K. Finally, one can use a Huffman code for example to encode these run lengths.

3.4 Exercises

3.1. An information source has an alphabet $\{u_1,u_2,u_3\}$ and generates a first-order Markov chain. The transition probabilities are given as follows:

$$
\begin{array}{lll}
P(u_1/u_1) = 1/2, & P(u_2/u_1) = 1/2, & P(u_3/u_1) = 0, \\
P(u_1/u_2) = 1/3, & P(u_2/u_2) = 0, & P(u_3/u_2) = 2/3, \\
P(u_1/u_3) = 1/3, & P(u_2/u_3) = 2/3, & P(u_3/u_3) = 0.
\end{array}
$$

a. Sketch the state diagram belonging to this Markov chain. Is the Markov chain ergodic?

b. Determine the probabilities of the symbols u_1, u_2 and u_3.

3.2. An information source with alphabet $\{0,1\}$ generates a second-order Markov chain, which is described by the following transition probabilities:

$$P(0/00) = 0.8, \qquad P(0/11) = 0.2,$$
$$P(1/00) = 0.2, \qquad P(1/11) = 0.8,$$
$$P(0/01) = 0.5, \qquad P(0/10) = 0.5,$$
$$P(1/01) = 0.5, \qquad P(1/10) = 0.5.$$

a. Sketch the state diagram belonging to this chain.
b. Calculate the probabilities of the states S_i.

3.3. An information source with source alphabet $\{0,1\}$ generates a second-order Markov chain which is described by the following transition probabilities:

$$P(0/00) = 1/4, \qquad P(0/01) = 1/4,$$
$$P(0/10) = 3/4, \qquad P(0/11) = 3/4.$$

a. Show that the Markov chain is completely described by the given transition probabilities.
b. Sketch the state diagram.
c. Calculate the probabilities of each of the states.
d. Calculate the probabilities of the source output symbols.
e. Indicate how the chain can be made non-ergodic by altering one of the given transition probabilities.

3.4. An information source has an alphabet $\{u_1,u_2,u_3\}$ and generates symbols which form a stationary first-order Markov chain. The transition probabilities are as follows:

$$P(u_1/u_1) = 0, \qquad P(u_2/u_1) = 1/5, \qquad P(u_3/u_1) = 4/5,$$
$$P(u_1/u_2) = 1/2, \qquad P(u_2/u_2) = 1/10, \qquad P(u_3/u_2) = 2/5,$$
$$P(u_1/u_3) = 1/2, \qquad P(u_2/u_3) = 0, \qquad P(u_3/u_3) = 1/2.$$

a. Sketch the state diagram.
b. Calculate the probabilities of the symbols u_1, u_2 and u_3.
c. Calculate the amount of information of a source without memory of which the symbols have the same probabilities as those of the source being considered.
d. Calculate the amount of information accompanying an arbitrary transition of the source with memory.
e. Calculate the joint amount of information of two symbols.

f. Calculate the redundancy, the dependence redundancy and the total redundancy.

g. Give a relation for the total redundancy in terms of the number of most probable messages.

3.5. An information source has an alphabet $\{u_1, u_2, u_3\}$ and generates a stationary first-order Markov chain. The transition probabilities of a symbol u_i to a symbol u_j with $i \neq j$ are all equal to $p/2$.

a. Sketch the state diagram belonging to this Markov chain.

b. Determine the probabilities of the symbols u_1, u_2 and u_3.

c. Calculate the amount of information with respect to an arbitrary transition.

d. Determine for what p this amount of information achieves a maximum.

e. What meaning do you attach to the values of $H(U)$ found for $p = 0$ and $p = 1$, and to the maximum value of $H(U)$?

3.6. An information source generates two symbols 0 and 1. The sequence of generated symbols forms an ergodic second-order Markov chain with transition probabilities

$$P(0/00) = 0.8,$$
$$P(0/01) = 0.5,$$
$$P(0/10) = 0.5,$$
$$P(0/11) = 0.2.$$

a. How large is the amount of information of a trigram originating from this information source? Use

$$P(00) = P(11) = 5/14,$$
$$P(01) = P(10) = 1/7.$$

Using this result, determine the amount of information per symbol, denoted by $H_3(U)$.

b. How large is the amount of information of a bigram? Hence determine $H_2(U)$.

c. How large is $H_1(U)$?

d. The conditional amount of information in the event that $N-1$ preceding symbols are known is denoted by $F_N(U)$. Determine $F_1(U)$, $F_2(U)$, $F_3(U)$.

e. Explain why $F_3(U) < F_2(U) < F_1(U)$.

f. What can you say about the values of $F_4(U)$ and $H_4(U)$?

g. Sketch $F_N(U)$ and $H_N(U)$ as a function of the number of symbols N.

3.5 Solutions

3.1. *a.* The Markov chain is of order $k = 1$, so that the number of states is $m^k = 3$. The state diagram then has the form shown in Figure 3.7.

Figure 3.7. State diagram of Exercise 3.1.

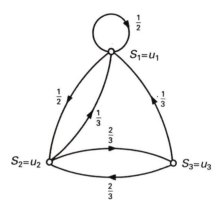

The chain is ergodic since from each state one can reach any other state.

b. Calculating the probabilities of the symbols u_1, u_2 and u_3 corresponds with calculating the probabilities of the three states S_1, S_2 and S_3, which follow from the equations

$$P(u_1) = P(u_1) \cdot P(u_1/u_1) + P(u_2) \cdot P(u_1/u_2) + P(u_3) \cdot P(u_1/u_3),$$

$$P(u_2) = P(u_1) \cdot P(u_2/u_1) + P(u_2) \cdot P(u_2/u_2) + P(u_3) \cdot P(u_2/u_3),$$

$$P(u_3) = P(u_1) \cdot P(u_3/u_1) + P(u_2) \cdot P(u_3/u_2) + P(u_3) \cdot P(u_3/u_3).$$

Furthermore, the following requirement must also be satisfied:

$$P(u_1) + P(u_2) + P(u_3) = 1.$$

Filling the given probabilities into these equations leads to

$$P(u_1) = P(u_1)\tfrac{1}{2} \quad + \tfrac{1}{3}P(u_2) \quad + \tfrac{1}{3}P(u_3),$$

$$P(u_2) = P(u_1)\tfrac{1}{2} \quad\quad\quad\quad\quad + \tfrac{2}{3}P(u_3),$$

$$P(u_3) = \quad\quad\quad\quad \tfrac{2}{3}P(u_2),$$

$$1 = P(u_1) \quad + P(u2) \quad + P(u_3).$$

Solving these four equations with three unknowns gives

$$P(u_1) = 10/25,$$

$P(u_2) = 9/25,$

$P(u_3) = 6/25.$

3.2. The Markov chain is of order $k = 2$, so that the chain can find itself in $m^k = 2^2 = 4$ different states. The state diagram is given in Figure 3.8.

b. The probabilities of the four states S_1, S_2, S_3 and S_4 can be calculated with the help of the following equations:

$$P(00) = P(00){\cdot}P(0/00) + P(10){\cdot}P(0/10),$$

$$P(01) = P(00){\cdot}P(1/00) + P(10){\cdot}P(1/10),$$

$$P(10) = P(01){\cdot}P(0/01) + P(11){\cdot}P(0/11),$$

$$P(11) = P(11){\cdot}P(1/11) + P(01){\cdot}P(1/01)$$

and

$$P(00) + P(01) + P(10) + P(11) = 1.$$

Filling in the given probabilities, this leads to

$$P(00) = P(00){\cdot}0.8 + P(10){\cdot}0.5,$$

Figure 3.8. State diagram of Exercise 3.2.

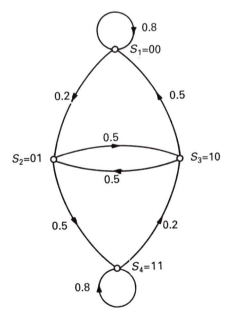

$$P(01) = P(00)\cdot 0.2 + P(10)\cdot 0.5$$
$$P(10) = P(01)\cdot 0.5 + P(11)\cdot 0.2$$
$$P(11) = P(01)\cdot 0.5 + P(11)\cdot 0.8$$
$$P(00) + P(01) + P(10) + P(11) = 1.$$

Solving these four equations with three unknowns leads to

$$P(00) = P(11) = 5/14,$$
$$P(01) = P(10) = 2/14.$$

3.3. *a.* The Markov chain is of order $k = 2$, so that the chain can find itself in $m^k = 2^2 = 4$ states, designated by 00, 01, 10 and 11. There are $m^{k+1} = 2^{2+1} = 8$ transitions and therefore also 8 transition probabilities. Of these, there are $m^{k+1} - m^k = 2^{2+1} - 2^2 = 4$ which can be chosen freely. These probabilities are given. The rest are fixed: $P(1/00) = 1 - P(0/00) = \frac{3}{4}$, etcetera.
b. The state diagram has the form of Figure 3.9.
c. The probabilities of the states are given by the relations

Figure 3.9. State diagram of Exercise 3.3.

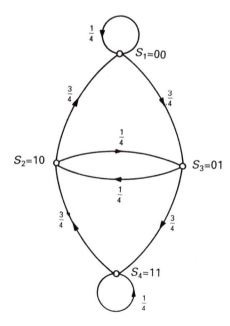

$$P(00) = \tfrac{1}{4}P(00) + \tfrac{3}{4}P(10),$$

$$P(01) = \tfrac{3}{4}P(00) + \tfrac{1}{4}P(10),$$

$$P(10) = \tfrac{1}{4}P(01) + \tfrac{3}{4}P(11),$$

$$P(11) = \tfrac{3}{4}P(01) + \tfrac{1}{4}P(11)$$

and

$$P(00) + P(10) + P(01) + P(11) = 1.$$

Hence

$$P(00) = P(01) = P(10) = P(11) = \tfrac{1}{4}.$$

d. The probability of a zero is now found from

$$P(0) =$$

$$P(00){\cdot}P(0/00) + P(01){\cdot}P(0/01) + P(10){\cdot}P(0/10) + P(11){\cdot}P(0/11)$$

$$= \tfrac{1}{4}\cdot\tfrac{1}{4} + \tfrac{1}{4}\cdot\tfrac{1}{4} + \tfrac{1}{4}\cdot\tfrac{3}{4} + \tfrac{1}{4}\cdot\tfrac{3}{4} = \tfrac{1}{2},$$

and therefore also

$$P(1) = \tfrac{1}{2}.$$

Figure 3.10. State diagram of Exercise 3.3.

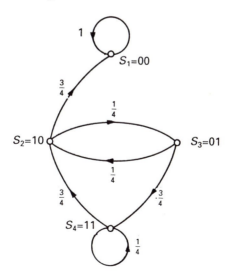

e. One way of making the chain non-ergodic is by creating two isolated parts. One can achieve this by choosing

$$P(0/00) = 1,$$

or

$$P(0/11) = 0$$

(so that $P(1/00) = 0$ and $P(1/11) = 1$). The choice $P(0/00) = 1$ results in the non-ergodic chain of Figure 3.10.

3.4. *a.* The state diagram has the form shown in Figure 3.11.

Figure 3.11. State diagram of Exercise 3.4.

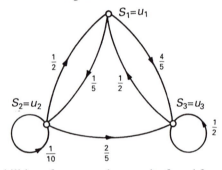

b. The probabilities of u_1, u_2 and u_3 can be found from the equations

$$P(u_1) = P(u_1) \cdot P(u_1/u_1) + P(u_2) \cdot P(u_1/u_2) + P(u_3) \cdot P(u_1/u_3),$$

$$P(u_2) = P(u_1) \cdot P(u_2/u_1) + P(u_2) \cdot P(u_2/u_2) + P(u_3) \cdot P(u_2/u_3),$$

$$P(u_3) = P(u_1) \cdot P(u_3/u_1) + P(u_2) \cdot P(u_3/u_2) + P(u_3) \cdot P(u_3/u_3)$$

and the requirement

$$P(u_1) + P(u_2) + P(u_3) = 1.$$

Filling in the given values gives

$$P(u_1) = 0P(u_1) \quad + \tfrac{1}{2}P(u_2) \qquad + \tfrac{1}{2}P(u_3),$$

$$P(u_2) = \tfrac{1}{5}P(u_1) \quad + \tfrac{1}{10}P(u_2) \qquad + 0P(u_3),$$

$$P(u_3) = \tfrac{4}{5}P(u_1) \quad + \tfrac{2}{5}P(u_2) \qquad + \tfrac{1}{2}P(u_3),$$

$$P(u_1) + P(u_2) + P(u_3) = 1.$$

From these four equations with three unknowns it then follows that

$$P(u_1) = 1/3, \qquad P(u_2) = 2/27, \qquad P(u_3) = 16/27.$$

c. The amount of information of a source without memory with the above probabilities for the symbols is found from

$$H(U) = -\sum_{i=1}^{3} P(u_i) \log P(u_i)$$

$$= -\frac{1}{3} \log \frac{1}{3} - \frac{2}{27} \log \frac{2}{27} - \frac{16}{27} \log \frac{16}{27} = 1.25 \quad \text{bits/symbol.}$$

d. For the amount of information of the source with memory for an arbitrary transition, we have per definition that

$$H(U_2/U_1) = -\sum_{i=1}^{3} \sum_{j=1}^{3} P(u_i) \cdot P(u_j/u_i) \log P(u_j/u_i)$$

$$= \sum_{i=1}^{3} P(u_i) \left[-\sum_{j=1}^{3} P(u_j/u_i) \log P(u_j/u_i) \right].$$

Substitution of the given probabilities yields

$$H(U_2/U_1) = \frac{1}{3} \left[0 - \frac{1}{5} \log \frac{1}{5} - \frac{4}{5} \log \frac{4}{5} \right]$$

$$+ \frac{2}{27} \left[-\frac{1}{2} \log \frac{1}{2} - \frac{1}{10} \log \frac{1}{10} - \frac{2}{5} \log \frac{2}{5} \right]$$

$$+ \frac{16}{27} \left[-\frac{1}{2} \log \frac{1}{2} - 0 - \frac{1}{2} \log \frac{1}{2} \right]$$

$$= 0.93 \quad \text{bit/symbol.}$$

e. From the relation

$$H(U_1, U_2) = H(U_1) + H(U_2/U_1),$$

and the fact that it may be assumed that the source is stationary, that is

$$H(U_1) = H(U_2) = H(U),$$

there follows

$$H(U_1, U_2) = H(U_1) + H(U_2/U_1) = 1.25 + 0.93 = 2.18 \text{ bits/message.}$$

f. The redundancy is defined as

$$\text{red} = 1 - \frac{H(U)}{\max_u H(U)} = 1 - \frac{H(U)}{\log n}.$$

Filling in the values gives

$$\text{red} = 1 - \frac{1.25}{\log 3} = 1 - \frac{1.25}{1.58} = 0.21.$$

For the independence redundancy we have

$$\text{red}_\infty = 1 - \frac{H_\infty(U)}{H(U)} = 1 - \frac{H(U_2/U_1)}{H(U)} = 1 - \frac{0.93}{1.25} = 0.26.$$

The total redundancy is a combination of both measures and is given by

$$\text{red}_{tot} = 1 - \frac{H_\infty(U)}{\max\limits_u H(U)} = 1 - \frac{0.93}{1.58} = 0.41.$$

g. From

$$M_{max} = 2^{l \cdot \max H(U)} \quad \text{there follows} \quad \max_u H(U) = \frac{1}{l} \log M_{max},$$

and from

$$M_\infty = 2^{l \cdot H_\infty(U)} \quad \text{there follows} \quad H_\infty(U) = \frac{1}{l} \log M_\infty.$$

Substitution of the expressions obtained for max $H(U)$ and $H_\infty(U)$ in the expression for the total redundancy gives

$$\text{red}_{tot} = 1 - \frac{H_\infty(U)}{\max\limits_u H(U)} = 1 - \frac{\log M_\infty}{\log M_{max}}.$$

3.5. *a.* The state diagram has the form of Figure 3.12.

Figure 3.12. State diagram of Exercise 3.5.

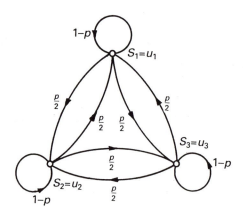

b. The probabilities of u_1, u_2 and u_3 follow from the equations

$$P(u_1) = P(u_1) \cdot P(u_1/u_1) + P(u_2) \cdot P(u_1/u_2) + P(u_3) \cdot P(u_1/u_3),$$

$$P(u_2) = P(u_1) \cdot P(u_2/u_1) + P(u_2) \cdot P(u_2/u_2) + P(u_3) \cdot P(u_2/u_3),$$

$$P(u_3) = P(u_1) \cdot P(u_3/u_1) + P(u_2) \cdot P(u_3/u_2) + P(u_3) \cdot P(u_3/u_3)$$

and

$$P(u_1) + P(u_2) + P(u_3) = 1.$$

By filling in the given values it follows that

$$P(u_1) = P(u_2) = P(u_3) = \frac{1}{3}.$$

The symmetry of the state diagram already suggests that each state will occur with equal probability, so that this calculated result agrees with the expectations.

c. For the amount of information with respect to an arbitrary transition we have that

Figure 3.13. Amount of information for arbitrary transition as function of p.

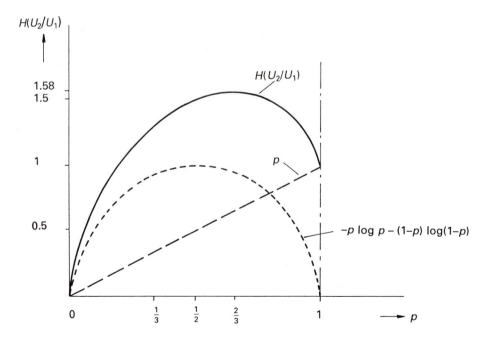

$$H(U_2/U_1) = -\sum_{i=1}^{3} \sum_{j=1}^{3} P(u_i) \, P(u_j/u_i) \log P(u_j/u_i)$$

$$= \sum_{i=1}^{3} P(u_i) \left[-\sum_{j=1}^{3} P(u_j/u_i) \log P(u_j/u_i) \right].$$

Substitution of the given values yields

$$H(U_2/U_1) = 3 \cdot \frac{1}{3} \left[-2 \frac{p}{2} \log \frac{p}{2} - (1-p) \log (1-p) \right]$$

$$= p - p \log p - (1-p) \log(1-p) \quad \text{bits/symbol.}$$

See Figure 3.13.

d. $H(U_2/U_1)$ is maximum if

$$\frac{\mathrm{d} H(U_2/U_1)}{\mathrm{d} p} = 0.$$

That is

$$1 - \frac{1}{\ln 2} - \log p + \frac{1}{\ln 2} + \log(1-p) = 0.$$

This gives $p = \frac{2}{3}$, from which it follows that $\max_u H(U) = \log 3 = 1.58$ bits/symbol.

e. If $p = 0$ the chain remains in the same state. There is therefore no longer any uncertainty, thus the amount of information is then 0. If $p = 1$ the chain has an equal chance of coming into one of both other states from any given state. Each symbol thus delivers 1 bit of information. For $p = \frac{2}{3}$ (maximum value of $H(U_2/U_1)$) every transition has the same probability of occurring. The chain thus behaves as if there are three independent symbols: if a previous symbol is given, the uncertainty is not reduced. Thus

$$H(U_2/U_1) = \log n = \log 3 \quad \text{bits/symbol.}$$

3.6. a. To calculate the amount of information per trigram the probabilities of the trigrams are determined first. These are

$$P(000) = 5/14 \times 8/10 = 4/14,$$
$$P(001) = 5/14 \times 2/10 = 1/14,$$
$$P(010) = 1/7 \times 1/2 = 1/14,$$
$$P(011) = 1/7 \times 1/2 = 1/14,$$
$$P(100) = 1/7 \times 1/2 = 1/14,$$
$$P(101) = 1/7 \times 1/2 = 1/14,$$

$$P(110) = 5/14 \times 2/10 = 1/14,$$
$$P(111) = 5/14 \times 8/10 = 4/14.$$

Hence, it follows for H (trigram) that

$$H(\text{trigram}) = -2 \cdot \frac{4}{14} \log \frac{4}{14} - 6 \cdot \frac{1}{14} \log \frac{1}{14} = 2.67 \quad \text{bits/trigram}.$$

Per symbol, this yields

$$H_3(U) = \frac{1}{3} H(\text{trigram}) = 0.89 \quad \text{bits/symbol}.$$

b. The probabilities of the bigrams are identical to the probabilities of the states, so that

$$H(\text{bigram}) = -2 \cdot \frac{5}{14} \log \frac{5}{14} - 2 \cdot \frac{1}{7} \log \frac{1}{7} = 1.86 \quad \text{bits/bigram},$$

and

$$H_2(U) = \frac{1}{2} H(\text{bigram}) = 0.93 \quad \text{bit/symbol}.$$

c. The value of $H_1(U)$ is now calculated from the probabilities $P(0)$ and $P(1)$. These probabilities can be calculated from

$$P(0) = \sum_{i=1} P_i(\text{bigram}) \frac{n_{0,i}}{n_{0,i} + n_{1,i}},$$

where $n_{0,i}$ is the number of zeros in the ith bigram and $n_{1,i}$ the number of ones in the ith bigram. Remark: this formula is valid solely for equal lengths of the code words.

Thus

$$P(0) = \frac{5}{14} \cdot 1 + \frac{5}{14} \cdot 0 + \frac{1}{7} \cdot \frac{1}{2} + \frac{1}{7} \cdot \frac{1}{2} = \frac{1}{2},$$

and

$$P(1) = \frac{1}{2}.$$

This gives

$$H_1(U) = -\frac{1}{2} \log \frac{1}{2} - \frac{1}{2} \log \frac{1}{2} = 1 \quad \text{bit/symbol}.$$

d. It can be derived for $F_N(U)$ that (see Section 3.2)

$$N H_N(U) = (N - 1) H_{N-1}(U) + F_N(U).$$

Filling in the known values of $H_N(U)$ then results in

$$F_1(U) = H_1(U) = 1 \quad \text{bit/symbol,}$$

$$F_2(U) = 2H_2(U) - H_1(U) = 0.86 \quad \text{bit/symbol,}$$

$$F_3(U) = 3H_3(U) - 2H_2(U) = 0.81 \quad \text{bit/symbol.}$$

e. For an increasing value of N, $F_N(U)$ will be determined by transition probabilities of which the value is dependent on an increasing number of preceding symbols. The uncertainty therefore decreases, so that $F_N(U)$ will also decrease.

f. The value of $F_4(U)$ is equal to that of $F_3(U)$. The reason is that the Markov chain is of the second order. Only the two preceding symbols now have influence on the value of the transition probabilities, the rest no longer have any influence. This means that the value of $F_N(U)$ does not change any more for $N \geq 3$. The value of $H_4(U)$ is smaller than that of $H_3(U)$, because with the calculation of $H_N(U)$ the first two symbols of the message remain of influence. It is clear that this influence does decrease, however, so that $H_N(U)$ is a monotonic decreasing function of N, with a limit value

$$\lim_{N \to \infty} H_N(U) = F_3(U).$$

g. The sketch of Figure 3.14 gives the course of $F_N(u)$ and $H_N(u)$ as functions of N.

Figure 3.14. $F_N(U)$ and $H_N(U)$ as functions of N.

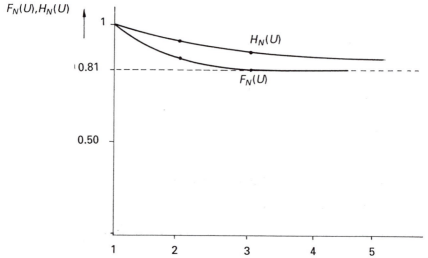

4

The discrete communication channel

4.1 Capacity of noiseless channels

The amount of information generated by an information source can be expressed in bits per symbol, as was often the case in the previous chapters. However, if the fact that an information source produces a sequence of symbols in a certain amount of time is also of importance, it may be advantageous to express the amount of information in bits per second. The relation between bits per symbol on the one side and bits per second on the other side is of course determined by the average duration of the symbols. When the amount of information produced by an information source has to be conveyed to a destination, one will have to make use of a communication channel that is able to transport the presented information. For the communication channel in this chapter we will assume the model given in Figure 4.1, which is similar to the model given in Figure 1.8.

Usually the information will be delivered by the source in a form that cannot directly be transported through the channel. The information must then be encoded into a code suitable for the channel. Furthermore, it may be desired to reduce the redundancy during the coding process, in order to limit as much as possible the amount of information that is to be conveyed through the channel; a problem which was studied in the foregoing chapters. With respect to the channel the obtained symbols are transformed to a signal

Figure 4.1. The discrete communication channel.

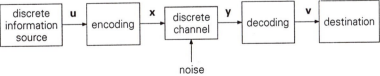

adapted to the physical properties of the channel, which may be distorted by noise during transport however. This has as consequence that a transmitted code symbol or message is not decoded as such after transmission through the channel, because symbols may have been altered through the influence of noise. Clearly from an information theoretic point of view, one is interested in the probability of an error occurring during the transmission of a symbol.

Besides this point, there is another aspect that deserves attention. In view of the physical properties of the channel, the amount of information to be conveyed will be finite. The information theorist is therefore also interested in the *capacity* of the communication model, this being the maximum amount of information that can be transported through the given channel.

Definition 4.1
The capacity C of a discrete noiseless channel is given by

$$C = \lim_{T \to \infty} \frac{\log N(T)}{T} \text{ bits/sec,} \tag{4.1}$$

where $N(T)$ is the number of allowed messages of duration T.   ○

From this definition it follows that for large T about 2^{CT} different messages can be transmitted through the channel during time T.

In the following the capacity of discrete noiseless channels whose symbols have not all the same duration or length will be given. Compare e.g. the Morse code, where a dot has a shorter duration than the dash.

As seen in Chapter 3 the structure of code words as a sequence of symbols can be considered as a result of a Markov process. Each symbol of a message can be considered as depending on its foregoing symbols. The states of the Markov process determine the succeeding symbol.

Let $N_j(T)$ be the number of possible different messages with duration T, ending in state S_j. Let t_{ij}^s be the duration of a symbol s whereby one goes from state S_i to S_j. Then $N_i(T - t_{ij}^s)$ is the number of messages ending with a transition from state S_i to state S_j by symbol s. Clearly, by summing over all states S_i and all symbols s leading to a transition to state S_j, the number $N_j(T)$ is found:

$$N_j(T) = \sum_i \sum_s N_i(T - t_{ij}^s). \tag{4.2}$$

By summing over all states S_j, $N(T)$ is obtained:

$$N(T) = \sum_j N_j(T). \tag{4.3}$$

By finding an expression for $N_j(T)$ and thus for $N(T)$, the capacity can be found.

Theorem 4.1
Let A be a matrix with elements

$$a_{ij} = \sum_s X^{-t_{ij}^s} - \delta_{ij}, \tag{4.4}$$

with δ_{ij} the Kronecker symbol ($\delta_{ij} = 1$ for $i = j$, otherwise $\delta_{ij} = 0$) and t_{ij}^s is the duration of a symbol s whereby one goes from state S_i to S_j.
The capacity of a discrete noiseless channel with symbols of unequal duration is given by

$$C = \log X_0, \tag{4.5}$$

whereby x_0 is the largest positive X for which the determinant of a matrix A is equal to 0: $|A| = 0$.

Proof
In fact the solution of

$$N_j(T) = \sum_i \left\{ \sum_s N_i(T - {}^s_{ij}) \right\} \tag{4.6}$$

is searched for, because then $N(T)$ can be computed, followed by the capacity with the help of Definition 4.1.
In fact equation (4.6) is a linear difference equation. For solving linear difference equations, methods have been derived which are in a way analogous to the methods for solving linear differential equations. Suppose that the relationship between a function $X(n)$ and $X(n-1)$ is given by

$$X(n) - \alpha X(n-1) = 0.$$

A particular solution is then given by

$$X(n) = c \lambda^n.$$

Filling this in the difference equation gives

$$c \lambda^n - \alpha c \lambda^{n-1} = 0.$$

And solving this characteristic equation gives

$$\lambda = \alpha,$$

and

$$X(n) = c \, \alpha^n.$$

For linear difference equations of higher order we find in a similar manner

$$X(n) - \alpha_1 X(n-1) - \alpha_2 X(n-2) - \ldots - \alpha_p X(n-p) = 0.$$

We now have p particular solutions of the form $c_k \lambda_k^n$, which when added together give the solution of the difference equation:

$$X(n) = c_1 \lambda_1^n + c_2 \lambda_2^n + \ldots + c_p \lambda_p^n.$$

Each particular solution, when filled in, gives a characteristic equation, for example

$$c_k \lambda_k^n - \alpha_1 c_k \lambda_k^{n-1} - \alpha_2 c_k \lambda_k^{n-2} - \ldots - \alpha_p c_k \lambda_k^{n-p} = 0,$$

from which λ_k may be determined.

If $\lambda_1 > \lambda_2 > \ldots > \lambda_p$ and $n \to \infty$, then in approximation $X(n) = c_1 \lambda_1^n$.

Note. Actually other terms, which have an oscillating character, must be added here for finite T.

The solution to equation (4.6) is now of the form

$$N_j(T) = \alpha_j X^T. \tag{4.7}$$

Substitution of this solution yields

$$\alpha_j X^T = \sum_i \sum_s \alpha_i X^{T - t_{ij}^s} \tag{4.8}$$

Dividing by X^T gives

$$\sum_i \sum_s \alpha_i X^{-t_{ij}^s} - \alpha_j = 0. \tag{4.9}$$

Write α_j as

$$\alpha_j = \sum_i \delta_{ij} \alpha_i,$$

with

and
$$\delta_{ij} = 1 \text{ for } i = j,$$

$$\delta_{ij} = 0 \text{ for } i \neq j.$$

We then get

$$\sum_i \alpha_i \left\{ \sum_s X^{-t_{ij}^s} - \delta_{ij} \right\} = 0. \tag{4.10}$$

The i functions between the braces are dependent. If we exclude the possibility of all α_i being zero, then the equality can only be satisfied if the wronskian determinant is equal to zero. Thus

$$\left[\begin{array}{ccc} \sum_s X^{-t_{11}^s} - 1 & \sum_s X^{-t_{12}^s} & \cdots \cdots \\ \sum_s X^{-t_{21}^s} & \sum_s X^{-t_{22}^s} - 1 & \cdots \cdots \\ \cdots & & \\ \cdots & & \end{array} \right] = 0.$$

In simplified form:

$$|A| = 0 \text{ with } a_{ij} = \sum_s X^{-t_{ij}^s} - \delta_{ij}. \tag{4.11}$$

The largest positive root X_0 of this funcion gives

$$N_j(T) = \alpha_j X_0^T, \tag{4.12}$$

and thus

$$N(T) = \sum_j \alpha_j X_0^T. \tag{4.13}$$

From this it follows for the capacity of the communication channel which makes use of this coding system that

$$C = \lim_{T \to \infty} \frac{\log N(T)}{T}$$

$$= \lim_{T \to \infty} \frac{\log \left(X_0^T \sum_j \alpha_j \right)}{T}$$

$$= \lim_{T \to \infty} \frac{T \log X_0}{T} + \lim_{T \to \infty} \left\{ \frac{\log \sum_s \alpha_j}{T} \right\}$$

$$= \log X_0. \tag{4.14}$$

□

In the following example we consider a Morse code and calculate the capacity.

Example 4.1
The Morse code has four symbols; a dot, a dash, a letter space and a word space, each with their own duration. See Figure 4.2. There is one restriction, namely that a letter space or a word space must always be followed by a dot or a dash. Spaces are therefore not allowed to follow each other. Hence we can discern two states: the state S1 after a space and the state S2 after a dot or a dash. This yields the state diagram given in Figure 4.3. For the elements of the matrix A we find

$$a_{11} = \sum_s X^{-t_{11}^s} - 1 = -1,$$

Figuur 4.2. Morse code.

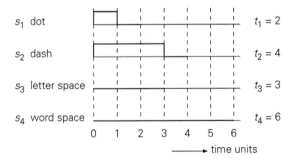

s_1 dot		$t_1 = 2$
s_2 dash		$t_2 = 4$
s_3 letter space		$t_3 = 3$
s_4 word space		$t_4 = 6$
	0 1 2 3 4 5 6	
	⟶ time units	

Figuur 4.3. State diagram with respect to Morse code.

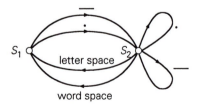

$$a_{12} = \sum_s X^{-t_{12}^s} \qquad = X^{-t_1} + X^{-t_2} \qquad = X^{-2} + X^{-4},$$

$$a_{21} = \sum_s X^{-t_{21}^s} \qquad = X^{-t_3} + X^{-t_4} \qquad = X^{-3} + X^{-6},$$

$$a_{22} = \sum_s X^{-t_{22}^s} - 1 \ = X^{-t_1} + X^{-t_2} - 1 \ = X^{-2} + X^{-4} - 1,$$

and thus the wronskian determinant leads to

$$|A| = \begin{bmatrix} -1 & X^{-2}+X^{-4} \\ X^{-3}+X^{-6} & X^{-2}+X^{-4}-1 \end{bmatrix} = 0,$$

or

$$f(X) = X^{-2} + X^{-4} + X^{-5} + X^{-7} + X^{-8} + X^{-10} - 1 = 0.$$

We find an approximation of the largest positive root of X graphically, after which a few calculations of $f(X)$ yield (see Figure 4.4)

$$X_0 \approx 1.454.$$

The capacity is therefore

$$C = \log X_0 = \log 1.454 = 0.54 \text{ bit/sec.}$$

If the restriction is dropped, and thus all symbols may follow one another, the system remains in one state. The diagram is now as given in Figure 4.5.

$$|A| = a_{11} = \sum_s X^{-t_{11}^s} - 1$$

Figure 4.4. Numerical approximation of the capacity by means of $f(X) = 0$.

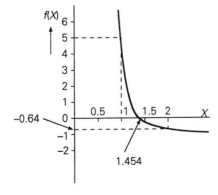

$$= X^{-t_1} + X^{-t_2} + X^{-t_3} + X^{-t_4} - 1$$

$$= X^{-2} + X^{-3} + X^{-4} + X^{-6} - 1 = 0.$$

Graphically we now find, in combination with a few calculations of $f(X)$,

$$X_0 \approx 1.51.$$

The capacity is now

$$C = \log 1.51 = 0.59 \text{ bit/sec.}$$

Thus with the help of the Morse code with and without the restriction, at most 0.54 and 0.59 bits per unit of time may be transported respectively. It depends on the manner in which the code is used whether or not this limit is reached.   △

4.2 Capacity of noisy channels

Before paying attention to such aspects as error probability and capacity, in the case of noisy channels the description of the communication channel will first be dealt with in more detail. We assume that, after coding and thus at the input of the communication channel (compare Figure 4.1), the symbols u_i delivered by the information source are encoded into code symbols x_i, belonging to the alphabet $X = \{x_1, x_2, \ldots, x_m\}$, whilst after having passed the channel code symbols y_j, belonging to an alphabet $Y = \{y_1, y_2, \ldots, y_n\}$, are received. The amount of information at the input can be denoted by $H(X)$ bits/second and at the output by $H(Y)$ bits/second. If we assume that a code word $\tilde{x}_i = (x_{i1}, x_{i2}, \ldots, x_{iN})$ with length L is presented and that as a result of this, a code word $\tilde{y}_j = (y_{j1}, y_{j2}, \ldots, y_{jN})$ is received, then the conditional probability of $\tilde{y}_j$ given $\tilde{x}_i$ can be given as follows:

Figure 4.5. State diagram in case of no restrictions.

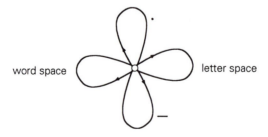

word space   letter space

$$q(\widetilde{y}_j/\widetilde{x}_i) = q(y_{j1}, y_{j2}, \ldots, y_{jL}/x_{i1}, \ldots x_{iL}). \tag{4.15}$$

In many cases one can assume that a received code symbol y_{jk} only depends on the transmitted symbol x_{ik} and not on the preceding symbols $x_{i1}, \ldots x_{i,k-1}$ or on the preceding symbols $y_{j1}, \ldots, y_{j,k-1}$. One then speaks of a *memoryless channel*. In that case, the transition probability of code word $\widetilde{x}_i$ to code word $\widetilde{y}_j$ can be split into the transition probabilities of the individual symbols, that is

$$q(\widetilde{y}_j/\widetilde{x}_i) = \prod_{k=1}^{L} q(y_{jk}/x_{ik}). \tag{4.16}$$

In future we can therefore determine the properties of the discrete channel, when this is assumed to be memoryless, per transmitted symbol instead of per code word, which considerably simplifies the analysis. The discrete channel can now be characterised by giving the matrix of conditional probabilities $q(y_j/x_i)$ for all symbols y_j given x_i. By using a shortened notation for the conditional probabilities, namely $q(y_j/x_i) = q_{ji}$, the so-called *channel matrix* receives the following shape:

$$Q = \begin{bmatrix} q_{11} & q_{21} & \cdots & q_{n1} \\ q_{12} & q_{22} & \cdots & q_{n2} \\ \vdots & \vdots & \vdots & \vdots \\ q_{1m} & q_{2m} & \cdots & q_{nm} \end{bmatrix}. \tag{4.17}$$

Because i denotes the input symbol and each input symbol has an output symbol as a result, the sum of each matrix row must be 1. The probability of an output symbol y_j occurring can be determined as follows:

$$q(y_j) = \sum_{i=1}^{m} p(x_i)\, q_{ji} \, ,$$

while the a posteriori probabilities of x_i are found by applying Bayes' theorem:

$$p(x_i/y_j) = \frac{p(x_i) \cdot q(y_j/x_i)}{q(y_j)} \, . \tag{4.18}$$

Using this, the conditional amounts of information $H(X/Y)$ and $H(Y/X)$ can be calculated according to Definition 1.3:

$$H(X/Y) = -\sum_{i=1}^{m} \sum_{j=1}^{n} q(y_j) \cdot p(x_i/y_j) \log p(x_i/y_j), \qquad (4.19)$$

and

$$H(Y/X) = -\sum_{i=1}^{m} \sum_{j=1}^{n} p(x_i) \cdot q(y_j/x_i) \log q(y_j/x_i). \qquad (4.20)$$

The conditional amount of information $H(X/Y)$, in other words the uncertainty about x if y has been received, is called the *equivocation*. The conditional amount of information $H(Y/X)$ can be viewed as the uncertainty in y if x is known. This uncertainty is the uncertainty that is introduced by the noise and is called the *irrelevance*.
If besides $H(X/Y)$ and $H(Y/X)$, the marginal amounts of information $H(X)$ and $H(Y)$ are also known, the amount of mutual information can be calculated on the basis of equation (1.36):

$$I(X;Y) = H(Y) - H(Y/X)$$

$$= H(X) - H(X/Y). \qquad (4.21)$$

From this last expression it follows that the mutual information can be regarded as the uncertainty at the receiving end about the transmitted symbol x before a symbol y has been received, reduced by the uncertainty that remains after a symbol y has been received. That is to say that $I(X;Y)$ is related to the amount of information that is transported over the channel. The notation R is sometimes also used instead of $I(X;Y)$ and we then speak of the *rate of transmission*. It can be expressed in bits/sec without loss of generality. For a non-distorted channel we clearly have that $H(X/Y) = H(Y/X) = 0$. In that case, $I(X;Y)$ will assume the maximum value $H(X)$. The transport of information can be depicted schematically in the manner indicated in Figure 4.6.
Note that only the transition probabilities $q(y_j/x_i)$ are fixed for a given channel. These transition probabilities thus represent, as it were, the influence of the noise. $I(X;Y)$ does not depend just on the probabilities $q(y_j/x_i)$, however, but also on the probabilities $p(x_i)$ of the input symbols, see equation (4.21), and can therefore still be varied for a given channel by varying the probabilities $p(x_i)$.
The most simple channel is a channel where messages are presented that are encoded in a binary alphabet $X = \{x_1, x_2\}$ and which result in messages with a binary alphabet $Y = \{y_1, y_2\}$ while, when transmitting x_1 and x_2, the probabilities of receiving y_1 and y_2 respectively are both equal to $1 - p$. Such

a channel is called a *binary symmetric channel, BSC.* The channel matrix for such a channel is given by

$$Q = \begin{bmatrix} 1-p & p \\ p & 1-p \end{bmatrix}.$$

By writing the input symbols beneath each other on the left and the output symbols on the right, and denoting the transitions with their probabilities with arrows, one can represent this channel as in Figure 4.7.

Example 4.2
The amount of transmitted information $I(X;Y)$ of a binary symmetric channel

Figure 4.6. Scheme for the transport of information.

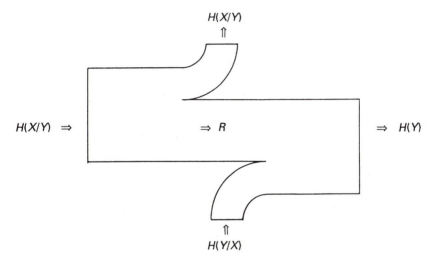

$H(X/Y) \Rightarrow$ $\Rightarrow R$ $\Rightarrow H(Y)$

Figure 4.7. The binary symmetric channel..

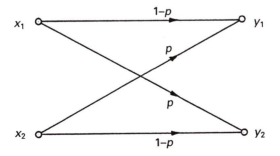

can be easily calculated. Suppose $p(x_1) = \alpha$ and therefore $p(x_2) = 1 - \alpha$, and assume $q(y_1) = \beta$ and $q(y_2) = 1 - \beta$. We then have for R

$$R = H(Y) - H(Y/X)$$

$$= -\beta \log \beta - (1-\beta) \log (1 - \beta) + p \log p + (1 - p) \log (1 - p)$$

with

$$\beta = \alpha(1 - p) + (1 - \alpha)p. \qquad\qquad \triangle$$

Besides binary symmetric channels, one can also distinguish non-binary and non-symmetric channels. Such channels will be treated in the following section.

In the foregoing we saw that from the point of view of the receiver, the uncertainty before a message has been received is equal to $H(X)$. After the receiver has received a message or symbol, the uncertainty about the transmitted symbol or message is reduced to $H(X/Y)$.

Thus in effect, an amount of information equal to $R = H(X) - H(X/Y)$ has been transported over the channel. The capacity C of a noisy channel is now defined as the maximum amount of information that can be transported over the channel. To this end, all possible information sources are connected to the channel so that all of the possible probability distributions $p(x_i)$ are traversed for $i = 1, \ldots, m$.

Definition 4.2
The capacity of a discrete noisy channel is given by

$$C = \max_{p(x)} R = \max_{p(x)} I(X;Y) = \max_{p(x)} \{H(Y) - H(Y/X)\}$$

$$= \max_{p(x)} \{H(X) - H(X/Y)\} \quad \text{bits/symbol.} \qquad (4.22)$$

$\bigcirc$

If the channel is not distorted, i.e. $H(X/Y) = H(Y/X) = 0$, C becomes the maximum value of $H(X)$. Since $\max\limits_{p(x)} H(X) = \log m$ the capacity then becomes

$$C = \log m \text{ bits/symbol.} \qquad (4.23)$$

If we assume that the symbols have a common duration of t seconds then the channel capacity per second is given by

$$C = \frac{1}{t} \log m \text{ bits/sec.} \qquad (4.24)$$

This corresponds to Definition 4.1 as can be seen as follows. If we consider code words of length L and with duration T, consisting of symbols of an alphabet with length m, then the number of possible code words equals $N(T) = m^L$. Because the symbols have an average duration t it should hold that $T = Lt$. Substitution of $N(T)$ and T in equation (4.1) of Definition 4.1 corresponds to equation (4.24). Thus, Definition 4.2 is a generalisation of Definition 4.1.

The capacity of a noisy channel is obviously less than the capacity of a noise free channel. If an amount of information $H(X)$ is presented, such that $H(X) < C$, then it is possible to transport this information with an arbitrarily small error ε. In the case where $H(X) > C$, it is possible to construct a code so that the channel can transport the information, but with the retention of an uncertainty $H(X/Y) = H(X) - C + \varepsilon$, with ε arbitrarily small. We will consider this point further in one of the following sections.

The calculation of the capacity in the case of noisy channels is generally no simple task and one will have to resort to numerical methods or to analytical approximations. For the previously mentioned binary symmetric channel, however, the calculation of the capacity is quite simple.

Example 4.3
Consider the binary symmetric channel as introduced in Example 4.2. It was found for the rate of transmission R that

$$R = -\beta \log \beta - (1 - \beta) \log (1 - \beta) + p \log p + (1 - p) \log (1 - p)$$

with $\beta = \alpha(1 - p) + (1 - \alpha)p$. In accordance with the definition of the capacity, the capacity of this channel is found by choosing that value of $p(x_1) = \alpha_0$ for which R is maximum. Thus

$$C = \max_{\alpha} R,$$

where R is given as above. As p is independent of α, R will be maximum if $-\beta \log \beta - (1 - \beta) \log (1 - \beta)$ is maximum. It can readily be seen on the basis of the properties of information measures that this is maximum if $\beta = \beta_0 = \frac{1}{2}$. It now follows that $\alpha_0 = \frac{1}{2}$, so that

$$C = \max_{\alpha} R = 1 + p \log p + (1 - p) \log (1 - p).$$

For $p = \frac{1}{2}$ we have that $C = 0$, so that no information is transported. For $p = 0$ or $p = 1$ on the other hand we have that $C = 1$, which means that $H(X)$ is completely transported. See Figure 4.8. $\triangle$

The capacity of a, not necessarily symmetric, binary channel can be simply determined in a graphical manner. The channel matrix of such a channel is

$$Q = \begin{bmatrix} q_{11} & q_{21} \\ q_{12} & q_{22} \end{bmatrix}.$$

Let $p(x_1) = \alpha$ and $q(y_1) = \beta$, and introduce

$$H_1 = -q_{11} \log q_{11} - q_{21} \log q_{21},$$

and

$$H_2 = -q_{12} \log q_{12} - q_{22} \log q_{22}.$$

Then

$$H(Y) = -\beta \log \beta - (1-\beta) \log (1-\beta)$$

and

$$H(Y/X) = \alpha H_1 + (1-\alpha) H_2 = H_2 + \alpha (H_1 - H_2).$$

Thus

$$R = H(Y) - H_2 - \alpha \cdot (H_1 - H_2).$$

Now

$$\beta = \alpha q_{11} + (1-\alpha)q_{12} = q_{12} + \alpha (q_{11} - q_{12}),$$

Or

$$\alpha = \frac{\beta - q_{12}}{q_{11} - q_{12}}.$$

In Figure 4.9 the length of DG represents the value of $H(Y)$, and we have that $AB = H_2$ and $KM = H_1$, thus $LM = H_1 - H_2$. It now follows that

Figure 4.8. Capacity for a binary symmetric channel.

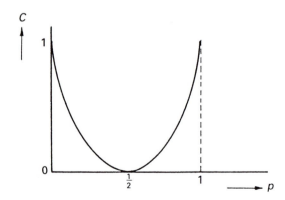

$$\frac{BE}{BL} = \frac{\beta - q_{12}}{q_{11} - q_{12}} = \alpha,$$

hence

$$EF = \alpha (H_1 - H_2),$$

so that FG represents the value of R. As the capacity is equal to the maximum value of R, we find R by drawing a line parallel to BM and tangent to the curve. This point of contact determines the value β_0 of β that belongs to the capacity. The optimal probability distribution of the source is therefore given by

$$p(x_1) = \frac{\beta_0 - q_{12}}{q_{11} - q_{12}} = \alpha_0 .$$

Usually in a binary channel it is assumed that $x_1 = y_1 = 0$ and $x_2 = y_2 = 1$. If the alphabet at the output is ternary, $Y = \{0,1,y\}$ for example, where y represents a symbol that is chosen if the receiver is not able to detect 0 or 1, then a *binary erasure channel (BEC)* arises. No interchange of the symbols 0 and 1 can occur in this case (see Figure 4.10(a), which depicts a symmetric *BEC*). Figure 4.10(b) represents an erasure channel where this is also possible: the so-called *errors-and-erasure channel*.
The channel matrix for the symmetric BEC reads

Figure 4.9. The capacity for a binary channel.

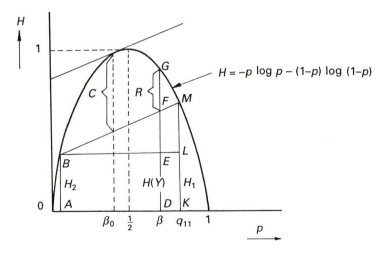

$$Q = \begin{bmatrix} 1-p & p & 0 \\ 0 & p & 1-p \end{bmatrix}.$$

The capacity of this channel is $1-p$ (see Figure 4.11 and also Exercise 4.7).

Theorem 4.2
The capacity of any discrete symmetric noisy channel is given by

$$C = \sum_{j=1}^{n} q(y_j/x_i) \log q(y_j/x_i) + \log n, \tag{4.25}$$

and is attained using the equiprobable distribution on the input alphabet.

Proof

$$C = \max_{p(x)} I(X;Y)$$

$$= \max_{p(x)} \left[\sum_{i=1}^{m} p(x_i) \sum_{j} q(y_j/x_i) \log q(y_j/x_i) - \sum_{j=1}^{n} q(y_j) \log q(y_j) \right]. \tag{4.26}$$

Figure 4.10. (a) Symmetric binary erasure channel; (b) errors-and-erasure channel.

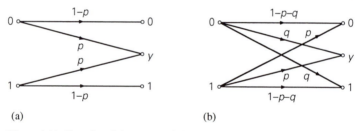

(a)   (b)

Figure 4.11. Capacity of the symmetric binary erasure channel.

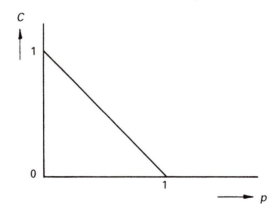

Because of the symmetry $\sum_j q(y_j/x_i) \log q(y_j/x_i)$ is identical for every i. Thus the first term is not influenced by the choice of the input probabilities. Hence

$$C = \sum_{j=1}^{n} q(y_j/x_i) \log q(y_j/x_i) - \max_{p(x)} \left[\sum_{j=1}^{n} q(y_j) \log q(y_j) \right]. \quad (4.27)$$

The second term can be achieved if the input probability is such that the output has the equiprobable distribution. Due to the symmetry of the transition matrix Q, this will be the case if the input probabilities are equiprobable. $\qquad\qquad\square$

As a final example we mention the *Z-channel*. For this model of a binary channel it is assumed that exactly one of the two input symbols, say x_1, is transmitted without error, while the other symbol may be incorrectly received with a probability p. The channel model is depicted in Figure 4.12. The channel matrix reads

$$Q = \begin{bmatrix} 1 & 0 \\ p & 1-p \end{bmatrix}.$$

Because this channel is not symmetric it will no longer hold that $p(x_1) = q(y_1)$ and $p(x_2) = q(y_2)$. Now we determine the capacity by supposing that $p(x_1) = 1 - \alpha_0$ and $p(x_2) = \alpha_0$, subsequently calculating the rate of transmission R and then determining the maximum of R. The calculation is given in Exercise 4.8. The capacity of the Z-channel is

$$C = \alpha_0 \left[p \log p - (1-p) \log \alpha_0 \right] - \left[1 - \alpha_0(1-p) \right] \log \left[1 - \alpha_0(1-p) \right], \quad (4.28)$$

where

$$\alpha_0 = \frac{1}{1-p + p^{-p/(1-p)}}. \quad (4.29)$$

As is apparent from this last channel model and Exercise 4.8, the calculation of the channel capacity for more complex channels quickly becomes

Figure 4.12. The Z-channel.

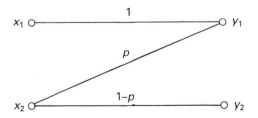

complex. One will then have to determine the capacity in a numerical way.

4.3 Error probability and equivocation

As has already been mentioned in Section 4.2, symbols may be altered during transportation over the communication channel due to the influence of noise, which results in errors. Besides the channel capacity, the quality of the communication channel will also be determined by the probability that an error is made during the transmission of a symbol. One can represent the (average) error probability in two manners, namely from the points of view of the receiver and of the transmitter. From now on we will assume that the number of symbols at the input of the channel is equal to the number of symbols at the output, so that we have a square channel matrix. If a symbol y_j has been received then an error is made if symbol x_i, $i \neq j$, is transmitted. This error probability is

$$p(e/y_j) = \sum_{\substack{i=1 \\ i \neq j}}^{n} p(x_i/y_j) = 1 - p(x_j/y_j). \tag{4.30}$$

Averaged out over all of the received symbols, this becomes

$$P_e = \sum_{j=1}^{n} q(y_j) \cdot p(e/y_j) = \sum_{j=1}^{n} q(y_j) \left[1 - p(x_j/y_j) \right]. \tag{4.31}$$

Similarly, from the point of view of the transmitter one can find as average error probability

$$P_e = \sum_{i=1}^{n} p(x_i) [1 - q(y_i/x_i)]. \tag{4.32}$$

That the average error probabilities are equal will be shown in the following theorem:

Theorem 4.3
For a communication channel with a square channel matrix, the average error probability from the point of view of the receiver is equal to that from the point of view of the transmitter.

Proof
In effect, it comes down to showing that the right-hand sides of equation (4.31) and equation (4.32) are equal to each other. With regard to equation (4.31) we have that

$$P_e = \sum_{j=1}^{n} q(y_j) \left[1 - p(x_j/y_j) \right] = \sum_{j=1}^{n} q(y_j) \left[1 - \frac{r(x_j,y_j)}{q(y_j)} \right]$$

$$= 1 - \sum_{j=1}^{n} r(x_j,y_j).$$

Since by definition

$$\sum_{i=1}^{n} \sum_{j=1}^{n} r(x_i,y_j) = 1,$$

it follows that

$$P_e = \sum_{i=1}^{n} \sum_{j=1}^{n} r(x_i,y_j) - \sum_{j=1}^{n} r(x_j,y_j) = \sum_{i=1}^{n} \sum_{\substack{j=1 \\ j \neq i}}^{n} r(x_i,y_j)$$

$$= \sum_{i=1}^{n} p(x_i) \sum_{\substack{j=1 \\ j \neq i}}^{n} q(y_j/x_i).$$

And finally

$$P_e = \sum_{i=1}^{n} p(x_i) \left[1 - q(y_i/x_i) \right].$$

This is same as formula (4.32). □

It is noted here that, as shown in the proof of Theorem 4.3 that the average error probability can also be written as

$$P_e = \sum_{i=1}^{n} \sum_{\substack{j=1 \\ j \neq i}}^{n} r(x_i,y_j). \tag{4.33}$$

A relationship exists between the equivocation $H(X/Y)$ and the average error probability P_e. This relationship is also known as *Fano's inequality*.

Theorem 4.4 (Fano's inequality)
Let $H(P_e)$ be defined by

$$H(P_e) = -P_e \log P_e - (1 - P_e) \log (1 - P_e),$$

where P_e is the average error probability, as given in equation (4.33). The following inequality now holds:

$$H(X/Y) \leq H(P_e) + P_e \log (n - 1). \tag{4.34}$$

Proof

The right-hand side of the inequality of formula (4.34) can be written with the help of equation (4.33) as

$$H(P_e) + P_e \log(n-1) = P_e \log\left(\frac{n-1}{P_e}\right) + (1 - P_e) \log\left(\frac{1}{1-P_e}\right)$$

$$= \sum_{\substack{i=1 \\ }}^{n} \sum_{\substack{j=1 \\ j\neq i}}^{n} r(x_i,y_j) \log\left(\frac{n-1}{P_e}\right) + \sum_{i=1}^{n} r(x_i,y_i) \log\left(\frac{1}{1-P_e}\right). \qquad (4.35)$$

The equivocation can be written in the same way in terms of similar summations.

$$H(X/Y) = -\sum_{i=1}^{n} \sum_{j=1}^{n} r(x_i,y_j) \log p(x_i/y_j)$$

$$= \sum_{\substack{i=1 \\ }}^{n} \sum_{\substack{j=1 \\ j\neq i}}^{n} r(x_i,y_j) \log\left(\frac{1}{p(x_i/y_j)}\right) + \sum_{i=1}^{n} r(x_i,y_i) \log\left(\frac{1}{p(x_i/y_i)}\right). \qquad (4.36)$$

Subtracting equation (4.35) from equation (4.36) we obtain

$$H(X/Y) - H(P_e) - P_e \log(n-1)$$

$$= \sum_{\substack{i=1 \\ }}^{n} \sum_{\substack{j=1 \\ j\neq i}}^{n} r(x_i,y_j) \log\left\{\frac{P_e}{(n-1)\,p(x_i/y_j)}\right\} + \sum_{i=1}^{n} r(x_i,y_i) \log\left\{\frac{1-P_e}{p(x_i/y_i)}\right\}. \qquad (4.37)$$

By going over to natural logarithms and making use of the inequality

$$\ln a \leq a - 1$$

it follows for the right-hand side of equation (4.37) that

$$\sum_{\substack{i=1 \\ }}^{n} \sum_{\substack{j=1 \\ j\neq i}}^{n} r(x_i,y_j) \log\left\{\frac{P_e}{(n-1)\,p(x_i/y_j)}\right\} + \sum_{i=1}^{n} r(x_i,y_i) \log\left\{\frac{1-P_e}{p(x_i/y_i)}\right\}$$

$$\leq \sum_{\substack{i=1 \\ }}^{n} \sum_{\substack{j=1 \\ j\neq i}}^{n} r(x_i,y_j) \frac{1}{\ln 2}\left\{\frac{P_e}{(n-1)\,p(x_i/y_j)} - 1\right\} + \sum_{i=1}^{n} r(x_i,y_i) \frac{1}{\ln 2}\left\{\frac{1-P_e}{p(x_i/y_j)} - 1\right\}$$

$$= \frac{1}{\ln 2} \left\{ \left[\frac{P_e}{n-1} \sum_{i=1}^{n} \sum_{\substack{j=1 \\ j \neq i}}^{n} \frac{r(x_i, y_j)}{p(x_i/y_j)} \right] - \sum_{i=1}^{n} \sum_{\substack{j=1 \\ j \neq i}}^{n} r(x_i, y_j) \right.$$

$$+ \sum_{i=1}^{n} \left\{ (1-P_e) \sum_{i=1}^{n} \frac{r(x_i, y_i)}{p(x_i/y_i)} \right\} - \sum_{i=1}^{n} r(x_i, y_i) \right\}$$

$$= \frac{1}{\ln 2} \left\{ \frac{P_e}{n-1} \cdot (n-1) - \left[1 - \sum_{i=1}^{n} r(x_i, y_i) \right] + (1-P_e) - \sum_{i=1}^{n} r(x_i, y_i) \right\} = 0.$$

From

$$H(X/Y) - H(P_e) - P_e \log (n-1) \leq 0,$$

formula (4.34) now follows directly.   □

In order to know under what condition equality occurs in Fano's inequality, we recall that the inequality $\ln a \leq a - 1$ becomes an equality if and only if $a = 1$. Without giving the proof the following requirements must then be satisfied:

$$p(x_i/y_j) = \frac{P_e}{n-1} \quad \text{for all } i \text{ and } j, \text{ with the exception of } i = j, \quad (4.38)$$

and

$$p(x_i/y_i) = 1 - P_e \quad \text{for all } i. \tag{4.39}$$

Since for all j

$$\sum_{i=1}^{n} p(x_i/y_j) = 1,$$

it can be derived that the condition mentioned in equation (4.39) is already implied by equation (4.38). Equation (4.38) implies that for a given output symbol, all input symbols have the same probability of occurrence with the exception of the one selected.

Besides the formal proof, Fano's inequality can be thought out as follows. The average uncertainty about X if Y is known can be regarded as the uncertainty of whether an error has been made or not, and if an error has been made (with probability P_e) the uncertainty of which of the $(n-1)$ remaining symbols has then been transmitted. The first amount of uncertainty is given by $H(P_e)$, while the second is at most equal to $\log (n-1)$, weighted with the probability P_e. From this inequality it follows that a small average error probability means that the equivocation is small. The inequality is depicted graphically in Figure 4.13.

4.4 Coding theorem for discrete memoryless channels

In case of a noisy channel the transmitted message can be affected by errors. Clearly, however, by transmitting the information in a redundant form the probability of error can be reduced, e.g. by repetition of the message. One would expect that in order to make the probability of error approach zero, the required redundancy is so large that the rate of transmission therefore also approaches zero. The characteristic of the following theorem is that it is possible to transmit information at the rate C through the channel with as small a probability of error or equivocation as desired.

This is not true for rates greater than C. If one tries to transmit at rate $C + R'$, then there will necessarily be an equivocation equal to or greater than R'.

Theorem 4.5 (Shannon's second coding theorem)
It is possible to transmit through a memoryless channel with capacity C an amount of information $H(X)$ with arbitrarily small probability of error (a small equivocation) if $H(X) \leq C$.

If $H(X) > C$ it is possible to encode the source in such a way that the equivocation is less than $H(X) - C + \varepsilon$, where ε is arbitrarily small. There is no coding method which gives an equivocation less than $H(X) - C$.

Proof
The theorem does not ask for the design of coding methods having the desired properties, but only the existence of such codes.

Now consider a source with amount of information $H(X)$. It is assumed that the probability distribution of the source is such that the capacity C is achieved. That means $C = H(X) - H(X/Y)$. The number of most probable

Figure 4.13. The relationship between error probability and equivocation.

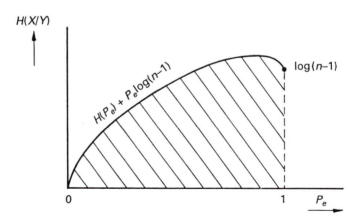

messages of length l is equal to $M_x = 2^{lH(X)}$; all these messages are equiprobable. The remaining messages have a small total probability. Similarly, the number of received messages with a high probability at the other side of the channel belong to a set with $M_y = 2^{lH(Y)}$ messages.

Each received message $\tilde{y}$ can come from a number of transmitted messages $\tilde{x}$ at the input, due to distortion in the noisy channel. The number of most probable inputs resulting in the same received message equals $M_{x/y} = 2^{lH(X/Y)}$.

In case of ideal adjustment of the source to the channel the number of transmitted probable messages is equal to $M_C = 2^{lC}$. In the case of non-ideal adjustment R < C, and thus $M_R = 2^{lR} < M_C$. We now consider a channel with R < C and assume the same information source.

We will associate the M_R messages with a selection of the possible channel inputs belonging to the set M_x and will show that it is possible to make the probability of error as small as possible.

The probability that a message $\tilde{x}_i \in M_x$ is transported is the probability that this message $\tilde{x}_i$ belongs to M_R. This probability is

$$P(\tilde{x}_i \in M_R) = \frac{2^{lR}}{2^{lH(X)}} = 2^{l\{R-H(X)\}},$$

which is (in approximation) equally large for all messages $\tilde{x}_i \in M_x$. We subsequently consider a received message $\tilde{y}_i$. The number of transmitted messages that on average can lead to the same received message $\tilde{y}_i$ is given by $M_{x/y}$. We now randomly choose a message $\tilde{x}_i$. An error can occur if besides $\tilde{x}_i$ there are one or more code words $\tilde{x}_j$ that belong to both $M_{x/y}$ and M_R and can therefore lead to the same received symbol $\tilde{y}_i$. The probability of an error is therefore

$$P_e = P \{\text{at least one } \tilde{x}_j, j \neq i, \in (M_{x/y} \cap M_R)\}. \tag{4.40}$$

This is the probability of the union of a number of events and it is known from probability theory that this is less than or equal to the sum of the probabilities of the individual events. From this it follows that

$$P_e \leq \sum_{\substack{j=1 \\ j \neq i}}^{M_{x/y}} P\left[(\tilde{x}_j \in M_{x/y}) \cap (\tilde{x}_j \in M_R)\right]. \tag{4.41}$$

The probability that a message belongs to $M_{x/y}$ and is also transmitted is

$$P\left[(\tilde{x}_j \in M_{x/y}) \cap (\tilde{x}_j \in M_R)\right] = P(\tilde{x}_j \in M_{x/y}) \cdot P(\tilde{x}_j \in M_R)$$

$$\leq P(\widetilde{x}_j \in M_R), \tag{4.42}$$

since by definition $P(\widetilde{x}_j \in M_{x/y}) \leq 1$. Hence

$$P_e \leq \sum_{\substack{j=1 \\ j \neq i}}^{M_{x/y}} P(\widetilde{x}_j \in M_R). \tag{4.43}$$

Substitution of the previously found value for $P(\widetilde{x}_j \in M_R)$ gives

$$P_e \leq \sum_{\substack{j=1 \\ j \neq i}}^{M_{x/y}} 2^{l\{R-H(X)\}},$$

that is

$$P_e \leq \{M_{x/y} - 1\}\, 2^{l\{R-H(X)\}}. \tag{4.44}$$

Since $R < C$, it can be established that

$$R = C - \theta = H(X) - H(X/Y) - \theta, \tag{4.45}$$

where θ is a positive constant. This gives

$$P_e \leq \{2^{lH(X/Y)} - 1\}2^{l(-H(X/Y)-\theta} \leq 2^{-l\theta}.$$

This means that the error probability can be made arbitrarily small for an increasing value of l, provided that $R < C$. Furthermore, the Fano inequality holds for the equivocation $H(X/Y)$:

$$H(X/Y) \leq H(P_e) + P_e \log(n-1).$$

Thus if P_e approaches zero the equivocation will also approach zero. On these grounds, one can infer that transmission is possible with a negligibly small error probability if $H(X) \leq C$.

The second part of the theorem can be shown as follows. If $H(X) > C$ the remainder of the information will be neglected. This gives at the receiver an equivocation $H(X/Y) > 0$. However, this equivocation will be at least $H(X) - C$. Assume the opposite, $H(X/Y) \leq H(X) - C$. Then $H(X/Y) = H(X) - C - \delta$ for some positive δ and thus

$$H(X) - H(X/Y) = C + \delta.$$

This contradicts the definition of C as the maximum of $H(X) - H(X/Y)$. Thus $H(X/Y) \geq H(X) - C$ and thus e.g. $H(X/Y) = H(X) - C + \varepsilon$ where ε can be arbitrarily small. $\qquad \square$

We note here that with Theorem 4.5 it has only been proven that error-free transmission is possible, but it is not indicated how this must be done. In practice, error-correcting codes are applied to reduce channel errors. Since one cannot take the lengths of the messages and accompanying code words to be infinitely large, for practical reasons, one will have to accept a certain error probability. Depending on the application this will usually vary from 10^{-3} to 10^{-12}.

4.5 Cascading of channels

In many cases the transport or storage of information will take place in such a way that the model used thus far, where we have just one channel, is too simple. We now consider a model involving cascading two or more channels. The information transmitted through each of these sub-channels is then equal to the input information lessened by the equivocation due to that sub-channel. We may now ask ourselves what relationship exists between the overall transmission rate and transmission rates per sub-channel. We suppose that X is the input of the first channel, while its output Y is again input for the second channel. The output of the second channel is given by Z. (Compare Figure 4.14.) An amount of information $H(X)$ is presented at the input, $H(Z)$ is received at the output, while $H(Y)$ passes between the two sections.

Theorem 4.6
If a channel with transmission rate R_1 is followed by a second channel then for the overall transmission rate R for both channels we have that

$$R \leq R_1. \tag{4.46}$$

Proof
We must prove that

$$R = H(X) - H(X/Z) \leq R_1 = H(X) - H(X/Y).$$

A symbol x_i from the alphabet X will have a symbol y_j from the alphabet Y as a result, which in turn has a symbol z_k from the alphabet Z as a result.

Figure 4.14. Cascading of channels.

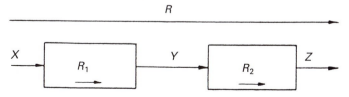

z_k thus depends solely on x_i via y_j, so that

$$p(z_k/y_j,x_i) = p(z_k/y_j),$$

for all i, j and k. By applying Bayes' theorem we find

$$\frac{p(x_i/y_j,z_k) \cdot p(z_k/y_j)}{p(x_i/y_j)} = p(z_k/y_j),$$

or

$$p(x/y,z) = p(x/y).$$

Now by using $\ln a \leq a - 1$ or $\log a \leq (a - 1) \log e$, we find

$$H(X/Z) - H(X/Y)$$

$$= -\sum_x \sum_z p(x,z) \log p(x/z) + \sum_x \sum_y p(x,y) \log p(x/y)$$

$$= -\sum_x \sum_y \sum_z p(x,y,z) \log p(x/z) + \sum_x \sum_y \sum_z p(x,y,z) \log p(x/y)$$

$$= -\sum_x \sum_y \sum_z p(x,y,z) \log \frac{p(x/z)}{p(x/y)} = -\sum_x \sum_y \sum_z p(x,y,z) \log \frac{p(x/z)}{p(x/y,z)}$$

$$= -\sum_y \sum_z p(y,z) \sum_x p(x/y,z) \log \frac{p(x/z)}{p(x/y,z)}$$

$$\geq -\sum_y \sum_z p(y,z) \log e \times \sum_x p(x/y,z) \left\{ \frac{p(x/z)}{p(x/y,z)} - 1 \right\}$$

$$= -\sum_y \sum_z p(y,z) \log e \times \left\{ \sum_x p(x/z) - \sum_x p(x/y,z) \right\}$$

$$= -\sum_y \sum_z p(y,z) \log e \times \{1 - 1\} = 0. \tag{4.47}$$

Hence it has been proved that $H(X/Z) \geq H(X/Y)$, from which it follows that $R \leq R_1$.   □

The theorem proven above is known as the *data processing theorem*. In actual fact the supposition indicates that only a loss of information is possible when processing data successively. One can retain all of the information only if there exists a unique relationship between input and output symbols. It is then a case of a channel with error-free transmission. In a more narrow sense the theorem indicates that cascading channels leads to a loss of source information because the equivocation increases. It is not possible to derive a simple relationship between the transmission rates R, R_1 and R_2.

Example 4.4
Consider two binary symmetric channels connected in cascade (Figure 4.15), where $0 \leq p \leq \frac{1}{2}$. In the case of equal probabilities for both input symbols we have for the first channel

$$R_1 = H(Y) - H(Y/X)$$

$$= 1 - H(P),$$

where

$$H(P) = -p \log p - (1-p) \log (1-p).$$

A symbol z_1 can arise from x_1 via y_1 or via y_2. Therefore

$$p(z_1/x_1) = (1-p)^2 + p^2 = 1 - 2p(1-p) = 1 - p',$$

and $\qquad p(z_2/x_1) = 2p(1-p) = p'.$

Clearly, it also holds that $p(z_1) = p(z_2) = \frac{1}{2}$. Then

$$R = H(Z) - H(Z/X)$$

$$= 1 - H(P')$$

where

$$H(P') = -p' \log p' - (1-p') \cdot \log (1-p').$$

For $0 \leq p \leq \frac{1}{2}$ it holds that $p \leq p' = 2p(1-p) \leq \frac{1}{2}$. Therefore, considering Figure 1.1, it may be concluded that

Figure 4.15. Cascade of two binary symmetric channels.

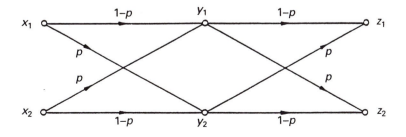

$$0 \leq H(P) \leq H(P') \leq 1$$

and thus

$$R \leq R_1.$$

At the same time it can be seen that $H(Z/X) = H(Z/Y) + H(Y/X)$ cannot hold. Because $H(Z/Y) = H(Y/X) = H(P)$ and $H(Z/X) = H(P')$, this would have to mean that $2H(P) = H(P')$, which, in general, is not true. Thus one cannot find the total equivocation from the sum of the sub-equivocations, but must determine the channel matrix of the whole channel, thus in this case the probabilities p' and $(1 - p')$. $\triangle$

4.6 Channels with memory

So far, we have considered channels without memory. This means that we assume that the occurrence of errors, that is to say the alteration of symbols during transportation, is independent of the preceding errors. Besides the advantage of a simple analysis of these channels, in practice it is a satisfactory model. Most of the error-correcting codes in use are therefore also based on the occurrence of independent errors. Yet a growing number of cases in practice deal with a channel with memory. This is due to the application of higher transmission rates through which imperfections during transmission can cause a sequence of successive errors. This effect also arises with the storage of data in digital optical and magnetic media such as disks and tapes because the occurrence of damage or defects can distort relatively more and more bits through the larger bit density per unit area.

In the case of channels which are not memoryless a stronger definition of the capacity is needed.

Assume input and output blocks of length L. Analogously to the capacity for single input symbols now we have

$$C = \lim_{L \to \infty} \frac{1}{L} \max I(X_1,\ldots,X_L;Y_1,\ldots,Y_L) \tag{4.48}$$

where the maximum is taken over all probability distributions on input blocks of length L. In general the evaluation of channels with memory is very difficult. We limit ourselves here to binary channels. The occurrence of a sequence of related errors is called a *burst*. It is not necessary that a sequence of consecutive errors occur; one will also often regard a temporary situation where a noticeably larger error probability is present as a burst. As the burst length one usually understands the length between the beginning of the first error and the end of the last error, without paying attention to the intervening errors. To examine this error behaviour one can use descriptive methods based on the statistical parameters measured from actually occurring errors, or construct generative models for the channels with memory that generate error sequences that are more or less similar to those of the actual channel. We will treat such a generative model here, which consists of a Markov chain with a certain number of states and the accompanying transition probabilities. The simplest model of this type is the model of Gilbert, which is depicted in Figure 4.16.

The Gilbert model of a channel with memory has two states G and B (good and burst) and generates a sequence of zeros and ones where a one represents an error and a zero therefore no error. An error bit is always a zero in state G (no error), while in state B an error bit is equal to zero with a probability of p and equal to one with a probability of $1 - p$. The Markov chain goes over into a new state every time an error is generated. The transition probabilities $P(G/B)$ and $P(B/G)$ are so small that the Markov chain has the tendency to remain in state G or B respectively. As is apparent from the model, a zero can be generated in both states G and B. To determine the probability of error, denoted here by $P(1)$, we must therefore introduce a state model as depicted in Figure 4.17 where now only state B_1 corresponds with an error and B_0 and G correspond with no occurrence of an error.

We can regard the transformation of the states G and B to the states in Figure 4.17 as a representation $f(.)$ such that

$$f(G) = 0, \qquad f(B_0) = 0, \qquad f(B_1) = 1.$$

For the Markov chain in Figure 4.16 it follows for the probabilities of the two states that

$$P(G) = \frac{P(G/B)}{P(B/G) + P(G/B)}$$

and

Figure 4.16. Model of Gilbert.

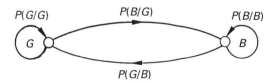

P(G/G)   P(B/G)   P(B/B)

G   B

P(G/B)

$$P(B) = \frac{P(B/G)}{P(B/G) + P(G/B)}.$$

An error occurs in state B with a probability of $(1 - p)$ so that we find

$$P(1) = (1 - p)\, P(B) = \frac{(1-p)\, P(B/G)}{P(B/G) + P(G/B)}.$$

The three probabilities $P(G/B)$, $P(B/G)$ and p of the Gilbert model therefore determine the probability $P(1)$ of the occurrence of errors. Although the Gilbert model itself is simple, the calculation of the detailed behaviour of the error sequences is rather complicated. Because it is a first-order Markov chain with only two states it has limitations, particularly with respect to the behaviour of the burst length, a reason why an extension to more states has been researched. Besides this, there is an extension where the probability of an error in the state G is not zero, but has a (small) positive value.

4.7 Exercises

4.1. Two different symbols x_1 and x_2 can be presented to a communication channel, with $p(x_1) = \frac{1}{2}$. The symbols y_1 and y_2 are received. One symbol is transmitted per second. The transition probabilities $q(y_j/x_i) = q_{ji}$ of this binary symmetric channel (BSC) are given by the following transition matrix:

$$Q = \begin{bmatrix} \frac{2}{3} & \frac{1}{3} \\ \frac{1}{3} & \frac{2}{3} \end{bmatrix}.$$

Figure 4.17. Model of Gilbert in expanded form.

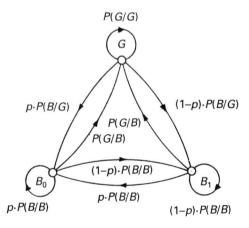

a. Give an expression for the average mutual information $I(X;Y)$, which only depends on the probability $p(x_i)$ of a symbol x_i and on the conditional probability $q(y_j/x_i)$ of a symbol y_j given x_i.

b. Calculate the mutual information $I(X;Y)$ with the help of the expression found in a.

c. Calculate the amount of information at the transmitting end.

d. Calculate the equivocation.

e. Calculate the mutual information $I(X;Y)$ by making use of the results from c and d.

4.2. Each second, one of two symbols x_i, $i = 1,2$, is presented to a communication channel with $p(x_1) = \alpha$. At the receiving end a choice can be made between 3 symbols y_j, $j=1,2,3$, where the transition probabilities $q(y_j/x_i) = q_{ji}$ are given by the following transition matrix:

$$Q = \begin{bmatrix} q_{11} & q_{21} & q_{31} \\ q_{12} & q_{22} & q_{32} \end{bmatrix} = \begin{bmatrix} \frac{1}{2} & \frac{1}{2} & 0 \\ \frac{1}{2} & \frac{1}{4} & \frac{1}{4} \end{bmatrix}.$$

a. Calculate the amount of information in the symbols at the receiving end.

b. Calculate the uncertainty $H(Y/X)$ due to the noise.

c. Calculate the capacity of this channel.

4.3. Three different symbols (x_1,x_2,x_3) can be submitted to a communication channel, each with a probability of $\frac{1}{3}$. Three different symbols (y_1,y_2,y_3) are observed at the receiving end. The transition probabilities $q(y_j/x_i) = q_{ij}$ are given by the following transition matrix:

$$Q = \begin{bmatrix} q_{11} & q_{21} & q_{31} \\ q_{12} & q_{22} & q_{31} \\ q_{13} & q_{23} & q_{33} \end{bmatrix} = \begin{bmatrix} \frac{10}{16} & \frac{2}{16} & \frac{4}{16} \\ \frac{5}{16} & \frac{6}{16} & \frac{5}{16} \\ \frac{6}{16} & \frac{1}{16} & \frac{9}{16} \end{bmatrix}.$$

a. How much information does one receive per symbol?

b. How large is the amount of information transported over this communication channel?

c. How large is the probability that x_2 has been transmitted, when y_1 is received?

4.4. An information source produces 3 symbols x_1, x_2 and x_3. This source is connected to a noisy channel. The receiver can distinguish 3 symbols y_1, y_2

and y_3. The above mentioned communication system is described by the matrix of joint probabilities $r(x_i,y_j) = r_{ij}$ with $i,j = 1,2,3$:

$$\begin{bmatrix} r_{11} & r_{12} & r_{13} \\ r_{21} & r_{22} & r_{23} \\ r_{31} & r_{32} & r_{33} \end{bmatrix} = \begin{bmatrix} \frac{1}{36} & \frac{1}{12} & 0 \\ \frac{1}{4} & \frac{1}{9} & \frac{5}{36} \\ 0 & \frac{1}{18} & \frac{1}{3} \end{bmatrix}.$$

With the help of the data given above, determine:

a. the amount of information in the symbols present at the receiving end;
b. the uncertainty $H(Y/X)$ due to noise;
c. the amount of information transported over this noisy channel.

4.5. A binary communication channel is made up of two cascading sub-channels (compare Figure 4.15).
The transition probabilities $q(z_j/x_i)$ for $i,j = 1,2,...$ of the first sub-channel are given by the transition matrix

$$Q_{Z/X} = \begin{bmatrix} \frac{3}{4} & \frac{1}{4} \\ \frac{1}{4} & \frac{3}{4} \end{bmatrix}.$$

The transition probabilities $q(y_j/z_i)$ for $i,j = 1,2,...$ of the second sub-channel are given by the transition matrix

$$Q_{Y/Z} = \begin{bmatrix} \frac{2}{3} & \frac{1}{3} \\ \frac{1}{3} & \frac{2}{3} \end{bmatrix}.$$

The probabilities of the symbols x_i, $i = 1,2,...$, are $\frac{1}{3}$ and $\frac{2}{3}$, and there is one symbol transmitted per second.

a. Determine $H(Z/X)$.
b. Determine the amount of information transported from X to Z, denoted by R_1.
c. Determine $H(Y/Z)$.
d. Determine the amount of information transported from X to Y, denoted by R.
e. Do you expect that R will be larger than R_1 or smaller? Why?

4.6. Three different symbols (x_1,x_2,x_3) can be presented to a communication channel, each with a probability of $\frac{1}{3}$. The symbols y_1, y_2 and y_3 are received.

If a symbol y_j is received then an error is made if a symbol x_i, $i{\neq}j$, has been transmitted. The transition probabilities $q(y_j/x_i) = q_{ji}$ are given by the following transition matrix:

$$Q = \begin{bmatrix} 0.5 & 0.3 & 0.2 \\ 0.4 & 0.3 & 0.3 \\ 0.1 & 0.9 & 0 \end{bmatrix}.$$

a. How much information $H(Y)$ does one receive per symbol at the receiving end?
b. What is the noise influence $H(Y/X)$?
c. Calculate the error probability $p(e/y_2)$ in the case when a symbol y_2 is received.
d. Calculate the average error probability from the point of view of the receiver.
e. Calculate the average error probability from the point of view of the sender. What do you notice?
f. Can you see directly from the transition matrix that the results of this channel are bad?
g. Calculate the amount of transmitted information and the equivocation.
h. Is the Fano inequality satisfied now?

4.7. The symbols of a binary information source with probabilities $p(x_1) = \alpha$ and $p(x_2) = 1 - \alpha$ are transmitted with the help of a binary erasure channel (BEC) with the following transition matrix:

$$Q = \begin{bmatrix} 1-p & p & 0 \\ 0 & p & 1-p \end{bmatrix}.$$

a. Calculate the uncertainty $H(Y/X)$ due to the influence of noise.
b. Determine the amount of transmitted information R.
c. Calculate the capacity C of this channel.

4.8. A binary source with probabilities $p(x_1) = (1-\alpha)$ and $p(x_2) = \alpha$ is connected to a Z-channel (see Figure 4.12).
a. Determine the amount of information $H(Y)$.
b. Calculate the amount of transmitted information R.
c. Determine the capacity C of this channel.

4.8 Solutions

4.1. *a*. The average mutual information $I(X;Y)$ of all possible symbols x and y is given by

$$
\begin{aligned}
I(X;Y) &= \sum_{i=j}^{2} \sum_{j=1}^{2} r(x_i,y_j) \log \frac{r(x_i,y_j)}{p(x_i)\, q(y_j)} \\
&= \sum_{i=j}^{2} \sum_{j=1}^{2} p(x_i)\, q(y_j/x_i) \log \frac{q(y_j/x_i)}{q(y_j)} \\
&= \sum_{i=j}^{2} \sum_{j=1}^{2} p(x_i)\, q(y_j/x_i) \log \frac{q(y_j/x_i)}{\displaystyle\sum_{k=1}^{2} p(x_k)\, q(y_j/x_k)} .
\end{aligned}
$$

b. Substitution of the given values in the above expression leads to

$$
\begin{aligned}
I(X;Y) = \ & \frac{1}{2}\left[\frac{2}{3}\log \frac{\frac{2}{3}}{\frac{1}{2}\cdot\frac{2}{3}+\frac{1}{2}\cdot\frac{1}{3}} + \frac{1}{3}\log \frac{\frac{1}{3}}{\frac{1}{2}\cdot\frac{2}{3}+\frac{1}{2}\cdot\frac{2}{3}} \right] \\
&+ \frac{1}{2}\left[\frac{1}{3}\log \frac{\frac{1}{3}}{\frac{1}{2}\cdot\frac{2}{3}+\frac{1}{2}\cdot\frac{1}{3}} + \frac{2}{3}\log \frac{\frac{2}{3}}{\frac{1}{2}\cdot\frac{1}{3}+\frac{1}{2}\cdot\frac{2}{3}} \right] \\
&= \frac{2}{3}\log\frac{4}{3}+\frac{1}{3}\log\frac{2}{3} = 0.08 \text{ bit/sec.}
\end{aligned}
$$

c. The amount of information $H(X)$ at the transmitting end is

$$
H(X) = -\frac{1}{2}\log\frac{1}{2} - \frac{1}{2}\log\frac{1}{2} = 1 \text{ bit/sec.}
$$

d. For the equivocation $H(X/Y)$ we have by definition

$$
H(X/Y) = -\sum_{i=1}^{2} \sum_{j=1}^{2} q(y_j)\, p(x_i/y_j) \log p(x_i/y_j).
$$

The probability of a symbol y_j is determined from

$$
q(y_j) = \sum_{i=1}^{2} p(x_i)\, q(y_j/x_i) = \sum_{i=1}^{2} p(x_i)\, q_{ji},
$$

therefore $\quad q(y_1) = \frac{1}{2}\cdot\frac{2}{3} + \frac{1}{2}\cdot\frac{1}{3} = \frac{1}{2},$

$$q(y_2) = \frac{1}{2} \cdot \frac{1}{3} + \frac{1}{2} \cdot \frac{2}{3} = \frac{1}{2},$$

while the a posteriori probabilities of x_i are found by applying the Bayes formula:

$$p(x_1/y_1) = \frac{p(x_1)\, q_{11}}{q(y_1)} = \frac{\frac{1}{2} \cdot \frac{2}{3}}{\frac{1}{2}} = \frac{2}{3},$$

$$p(x_2/y_1) = \frac{1}{3}, \qquad p(x_1/y_2) = \frac{1}{3}, \qquad p(x_2/y_2) = \frac{2}{3}.$$

Substitution of these values into the equation for the equivocation gives

$$H(X/Y) = -\frac{1}{2}\left[\frac{2}{3}\log\frac{2}{3} + \frac{1}{3}\log\frac{1}{3}\right] - \frac{1}{2}\left[\frac{1}{3}\log\frac{1}{3} + \frac{2}{3}\log\frac{2}{3}\right]$$

$$= -\frac{2}{3}\log\frac{2}{3} - \frac{1}{3}\log\frac{1}{3} = 0.91 \text{ bit/sec.}$$

e. The following relation holds for the mutual information:

$$I(X;Y) = H(X) - H(X/Y).$$

Substitution of the answers found in c and d gives

$$I(X;Y) = 1 - 0.91 = 0.09 \quad \text{bit/sec},$$

which is in agreement with the answer found in b.

4.2. *a.* The channel can be represented as shown in Figure 4.18.

Figure 4.18. Representation of the channel of Exercise 4.2.

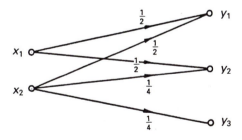

It is given that $p(x_1) = \alpha$, so that $p(x_2) = 1 - \alpha$. The probability of a symbol y_j occurring can now be determined with

$$q(y_j) = \sum_{i=1}^{2} p(x_i)\, q(y_j/x_i) = \sum_{i=1}^{2} p(x_i)\, q_{ji}.$$

This gives as result

$$q(y_1) = \frac{1}{2}\alpha + \frac{1}{2}(1 - \alpha) = \frac{1}{2},$$

$$q(y_2) = \frac{1}{2}\alpha + \frac{1}{4}(1 - \alpha) = \frac{1+\alpha}{4},$$

$$q(y_3) = 0\alpha + \frac{1}{4}(1 - \alpha) = \frac{1-\alpha}{4}.$$

Hence, for the received amount of information $H(Y)$

$$H(Y) = -\frac{1}{2}\log\frac{1}{2} - \frac{1+\alpha}{4}\log\frac{1+\alpha}{4} - \frac{1-\alpha}{4}\log\frac{1-\alpha}{4}$$

$$= \frac{3}{2} - \frac{1+\alpha}{4}\log(1 + \alpha) - \frac{1-\alpha}{4}\log(1 - \alpha) \quad \text{bits/sec.}$$

b. By definition it holds for $H(Y/X)$ that

$$H(Y/X) = -\sum_{i=1}^{2}\sum_{j=1}^{3} p(x_i) \cdot q(y_j/x_i) \cdot \log q(y_j/x_i)$$

$$= -\sum_{i=1}^{2}\sum_{j=1}^{3} p(x_i)\, q_{ji} \log q_{ji}$$

$$= -\alpha(\frac{1}{2}\log\frac{1}{2} + \frac{1}{2}\log\frac{1}{2}) - (1-\alpha)(\frac{1}{2}\log\frac{1}{2} + \frac{1}{4}\log\frac{1}{4} + \frac{1}{4}\log\frac{1}{4})$$

$$= \alpha + (1-\alpha)\cdot\frac{3}{2} = \frac{3}{2} - \frac{\alpha}{2} \quad \text{bits/sec.}$$

c. The amount of transported information follows from

$$R = H(Y) - H(Y/X) = \frac{3}{2} - \frac{1-\alpha}{4}\log(1 + \alpha) - \frac{1-\alpha}{4}\log(1 - \alpha) - \frac{3}{2} + \frac{\alpha}{2}$$

$$= -\frac{1+\alpha}{4}\log(1 + \alpha) - \frac{1-\alpha}{4}\log(1 - \alpha) + \frac{\alpha}{2} \quad \text{bits/sec.}$$

The capacity is found by choosing that value for $p(x_1) = \alpha$ for which R is maximum, that is to say

$$C = \max_{\alpha} R.$$

This value is found by solving the equation $dR/d\alpha = 0$:

$$\frac{dR}{d\alpha} = -\frac{1}{4}\log(1 + \alpha) - \frac{\log e}{4} + \frac{1}{4}\log(1 - \alpha) + \frac{\log e}{4} + \frac{1}{2} = 0.$$

Thus $\log\dfrac{1+\alpha}{1-\alpha} = 2.$

Hence

$\alpha = \dfrac{3}{5}$.

Substituting this in R gives

$C = 0.16$ bit/sec.

4.3. *a.* The channel can be represented as in Figure 4.19.

Figure 4.19. Representation of the channel of Exercise 4.3.

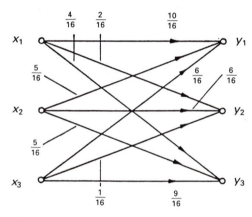

It is given that

$$p(x_1) = p(x_2) = p(x_3) = \tfrac{1}{3}.$$

The probabilities $q(y_j), j = 1,2,3$, can therefore be calculated as follows:

$$q(y_j) = \sum_{i=1}^{3} p(x_i)\, q(y_j/x_i) = \sum_{i=1}^{3} p(x_i)\, q_{ji}.$$

This gives as result

$$q(y_1) = \tfrac{1}{3} \left(\tfrac{10}{16} + \tfrac{5}{16} + \tfrac{6}{16} \right) = \tfrac{7}{16},$$

$$q(y_2) = \tfrac{1}{3} \left(\tfrac{2}{16} + \tfrac{6}{16} + \tfrac{1}{16} \right) = \tfrac{3}{16},$$

$$q(y_3) = \frac{1}{3} \left(\frac{4}{16} + \frac{5}{16} + \frac{9}{16} \right) = \frac{6}{16}.$$

Hence, for the received amount of information

$$H(Y) = -\frac{7}{16} \log \frac{7}{16} - \frac{3}{16} \log \frac{3}{16} - \frac{6}{16} \log \frac{6}{16} = 1.51 \text{ bits/symbol.}$$

b. The calculation of the transmitted amount of information is limited to the determination of the noise influence $H(Y/X)$. This is

$$H(Y/X) = -\sum_{i=1}^{3} \sum_{j=1}^{3} p(x_i) \, q(y_j/x_i) \log q(y_j/x_i) = -\sum_{i=1}^{3} \sum_{j=1}^{3} p(x_i) \, q_{ji} \log q_{ji}$$

$$= -\frac{1}{3} \left(\frac{10}{16} \log \frac{10}{16} + \frac{2}{16} \log \frac{2}{16} + \frac{4}{16} \log \frac{4}{16} \right) - \frac{1}{3} \left(\frac{5}{16} \log \frac{5}{16} + \frac{6}{16} \log \frac{6}{16} + \frac{5}{16} \log \frac{5}{16} \right)$$

$$- \frac{1}{3} \left(\frac{6}{16} \log \frac{6}{16} + \frac{1}{16} \log \frac{1}{16} + \frac{9}{16} \log \frac{9}{16} \right)$$

$$= -\frac{1}{3} \cdot \frac{1}{16} (10 \log 10 + 2 + 8 + 10 \log 5 + 12 \log 6 + 9 \log 9 - 48 \log 16).$$

By working this out one finds

$$H(Y/X) = 1.38 \text{ bits/symbol.}$$

The amount of transported information thus becomes

$$R = H(Y) - H(Y/X) = 1.51 - 1.38 = 0.13 \text{ bit/symbol.}$$

c. The probability that x_2 has been transmitted if y_1 has been received is the a posteriori probability of x_2. The probability $q(y_1/x_2)$ is known. Thus it seems obvious to use the Bayes formula. It then follows that

$$p(x_2/y_1) = \frac{p(x_2) \cdot q(y_1/x_2)}{q(y_1)} = \frac{\frac{1}{3} \cdot \frac{5}{16}}{\frac{7}{16}} = \frac{5}{21}.$$

4.4. a. To determine $H(Y)$, $q(y_j)$, $j = 1,2,3$, must first be calculated with the help of the matrix. We find these probabilities by summing the joint probabilities over all i:

$$\sum_{i=1}^{3} r(r_{ij}) = q(y_j).$$

Thus $q(y_1) = \frac{1}{36} + \frac{1}{4} = \frac{5}{18}$,

$$q(y_2) = \frac{1}{12} + \frac{1}{9} + \frac{1}{18} = \frac{1}{4},$$

$$q(y_3) = \frac{5}{36} + \frac{1}{3} = \frac{17}{36}.$$

Now

$$H(Y) = -\frac{5}{18} \log \frac{5}{18} - \frac{1}{4} \log \frac{1}{4} - \frac{17}{36} \log \frac{17}{36} = 1.52 \text{ bits/symbol.}$$

b. The uncertainty caused by noise is $H(Y/X)$. There are two methods of determining $H(Y/X)$. The first is to calculate $H(X,Y)$ from the given matrix and subtracting $H(X)$ from this; the second is to calculate $q(y_j/x_i)$ for $i,j =$ 1,2,3 and then to calculate $H(Y/X)$ directly from this. Probabilities $p(x_i)$ must be determined in both cases. The first method gives a little less calculation. By summing over j in the given matrix we find $p(x_i)$:

$$p(x_1) = \frac{1}{36} + \frac{1}{12} = \frac{1}{9},$$

$$p(x_2) = \frac{1}{4} + \frac{1}{9} + \frac{5}{36} = \frac{1}{2},$$

$$p(x_3) = \frac{1}{18} + \frac{1}{3} = \frac{7}{18},$$

so that

$$H(X) = -\frac{1}{9} \log \frac{1}{9} - \frac{1}{2} \log \frac{1}{2} - \frac{7}{18} \log \frac{7}{18} = 1.38 \text{ bits/symbol.}$$

Further,

$$H(X,Y) = -\frac{1}{36} \log \frac{1}{36} - \frac{1}{12} \log \frac{1}{12} - \frac{1}{4} \log \frac{1}{4} - \frac{1}{9} \log \frac{1}{9}$$

$$-\frac{5}{36} \log \frac{5}{36} - \frac{1}{18} \log \frac{1}{18} - \frac{1}{3} \log \frac{1}{3} = 2.45 \text{ bits/symbol,}$$

so that

$$H(Y/X) = H(X,Y) - H(X) = 2.45 - 1.38 = 1.07 \text{ bits/symbol.}$$

c. The amount of transmitted information R can be found from

$$\text{R} = H(Y) - H(Y/X) = 1.52 - 1.07 = 0.45 \text{ bit/symbol.}$$

4.5. *a.* The channel can be represented as in Figure 4.20.
The uncertainty due to the noise $H(Z/X)$ is calculated with the expression

$$H(Z/X) = -\sum_{i=1}^{2} \sum_{j=1}^{2} p(x_i)\, q(z_j/x_i)\, \log\, q(z_j/x_i)$$

$$= -\frac{1}{3}\left[\frac{3}{4}\log\frac{3}{4} + \frac{1}{4}\log\frac{1}{4}\right] - \frac{2}{3}\left[\frac{1}{4}\log\frac{1}{4} + \frac{3}{4}\log\frac{3}{4}\right]$$

$$= -\frac{3}{4}\log\frac{3}{4} - \frac{1}{4}\log\frac{1}{4} = 0.81 \ \ \text{bit/symbol.}$$

Figure 4.20. Cascading channels of Exercise 4.5.

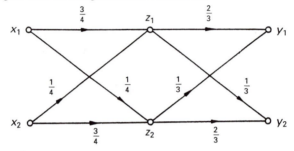

b. For R_1 we have

$$R_1 = H(Z) - H(Z/X).$$

$H(Z)$ must therefore be calculated first. Now

$$q(z_1) = p(x_1)\, q(z_1/x_1) + p(x_2)\, q(z_1/x_2) = \frac{1}{3}\cdot\frac{3}{4} + \frac{2}{3}\cdot\frac{1}{4} = \frac{5}{12},$$

and

$$q(z_2) = p(x_1)\, q(z_2/x_1) + p(x_2)\, q(z_2/x_2) = \frac{1}{3}\cdot\frac{1}{4} + \frac{2}{3}\cdot\frac{3}{4} = \frac{7}{12},$$

so that

$$H(Z) = -\frac{5}{12}\log\frac{5}{12} - \frac{7}{12}\log\frac{7}{12} = 0.97 \ \ \text{bit/symbol.}$$

This finally gives

$$R_1 = 0.97 - 0.81 = 0.16 \ \ \text{bit/sec.}$$

c. For $H(Y/Z)$ we have

$$H(Y/Z) = -\sum_{i=1}^{2} \sum_{j=1}^{2} p(z_1)\, p(y_j/z_i)\, \log\, p(y_j/z_i)$$

$$= -\frac{5}{12}\left[\frac{2}{3}\log\frac{2}{3} + \frac{1}{3}\log\frac{1}{3}\right] - \frac{7}{12}\left[\frac{2}{3}\log\frac{2}{3} + \frac{1}{3}\log\frac{1}{3}\right]$$

$$= -\frac{2}{3}\log\frac{2}{3} - \frac{1}{3}\log\frac{1}{3} = 0.92 \text{ bit/symbol.}$$

d. It follows for the probabilities at the output of the channel that

$$q(y_1) = q(z_1)\,q(y_1/z_1) + q(z_2)\,q(y_1/z_2) = \frac{5}{12}\cdot\frac{2}{3} + \frac{7}{12}\cdot\frac{1}{3} = \frac{17}{36},$$

$$q(y_2) = q(z_1)\,q(y_2/z_1) + q(z_2)\,q(y_2/z_2) = \frac{5}{12}\cdot\frac{1}{3} + \frac{7}{12}\cdot\frac{2}{3} = \frac{19}{36},$$

whence

$$H(Y) = -\frac{17}{36}\log\frac{17}{36} - \frac{19}{36}\log\frac{19}{36} = 0.99 \text{ bit/symbol.}$$

To determine $H(Y/X)$ one can imagine the two channels in series as being replaced by a new channel with input symbols x and output symbols y. The new transition matrix can then be found by determining along which routes one can come to the same output symbol from an input symbol and adding the corresponding probabilities together. Thus for example

$$q(y_1/x_1) = q(z_1/x_1)\,q(y_1/z_1) + q(z_2/x_1)\,q(y_1/z_2) = \frac{3}{4}\cdot\frac{2}{3} + \frac{1}{4}\cdot\frac{1}{3} = \frac{7}{12}.$$

The following matrix arises in this manner for the four transitions:

$$\begin{bmatrix} \frac{7}{12} & \frac{5}{12} \\ \frac{5}{12} & \frac{7}{12} \end{bmatrix}.$$

We have in actual fact just multiplied the matrices. The influence of the noise is now

$$H(Y/X) = -\frac{7}{12}\log\frac{7}{12} - \frac{5}{12}\log\frac{5}{12} = 0.97 \text{ bit/symbol.}$$

So that it follows that

$$R = H(Y) - H(Y/X) = 0.99 - 0.97 = 0.02 \text{ bit/sec.}$$

e. The uncertainty whether an input symbol does indeed result in a corresponding output symbol increases for two channels in series, because more alterations are now possible. This makes it more difficult to reconstruct which input symbol was the cause of an observed output symbol. Therefore

$$R \le R_1.$$

4.6. *a.* The amount of information $H(Y)$ can be found from the probabilities $q(y_1)$, $q(y_2)$ and $q(y_3)$. We have

$$q(y_j) = \sum_{i=1}^{3} p(x_i)\, q(y_j/x_i) = \sum_{i=1}^{3} p(x_i)\, q_{ji}.$$

Hence

$$q(y_1) = \frac{1}{3}\left(\frac{5}{10} + \frac{4}{10} + \frac{1}{10}\right) = \frac{1}{3},$$

$$q(y_2) = \frac{1}{3}\left(\frac{3}{10} + \frac{3}{10} + \frac{9}{10}\right) = \frac{1}{2},$$

$$q(y_3) = \frac{1}{3}\left(\frac{2}{10} + \frac{3}{10} + 0\right) = \frac{1}{6}.$$

Therefore

$$H(Y) = -\frac{1}{3}\log\frac{1}{3} - \frac{1}{2}\log\frac{1}{2} - \frac{1}{6}\log\frac{1}{6} = 1.46 \;\; \text{bits/symbol}.$$

b. For the noise influence we have

$$H(Y/X) = -\sum_{i=1}^{3}\sum_{j=1}^{3} p(x_i) \cdot q(y_j/x_i) \cdot \log q(y_j/x_i)$$

$$= -\frac{1}{3}\left[0.5 \log 0.5 + 0.3 \log 0.3 + 0.2 \log 0.2\right]$$

$$-\frac{1}{3}\left[0.4 \log 0.4 + 0.3 \log 0.3 + 0.3 \log 0.3\right]$$

$$-\frac{1}{3}\left[0.1 \log 0.1 + 0.9 \log 0.9\right]$$

$$= 0.495 + 0.523 + 0.158 = 1.18 \;\; \text{bits/symbol}.$$

c. If symbol y_2 has been received then an error has been made if symbol x_1 or x_3 was transmitted. This error probability is

$$p(e/y_2) = p(x_1/y_2) + p(x_3/y_2) = 1 - p(x_2/y_2).$$

With the help of the Bayes formula it directly follows that

$$p(x_2/y_2) = \frac{q(y_2/x_2) \cdot p(x_2)}{q(y_2)} = \frac{0.3 \cdot \frac{1}{3}}{\frac{1}{2}} = \frac{1}{5},$$

so that

$$p(e/y_2) = 1 - \frac{1}{5} = \frac{4}{5}.$$

d. It follows for the error probability P_e from the point of view of the receiver that

$$P_e = \sum_{j=1}^{3} q(y_j)\left[1 - p(x_j/y_j)\right].$$

With the help of the Bayes formula we find

$$p(x_1/y_1) = \frac{q(y_1/x_1) \cdot p(x_1)}{q(y_1)} = \frac{0.5 \frac{1}{3}}{\frac{1}{3}} = \frac{1}{2},$$

$$p(x_2/y_2) = \frac{1}{5},$$

$$p(x_3/y_3) = 0,$$

so that

$$P_e = \frac{1}{3}(1 - \frac{1}{2}) + \frac{1}{2}(1 - \frac{1}{5}) + \frac{1}{6}(1 - 0) = \frac{1}{6} + \frac{2}{5} + \frac{1}{6} = \frac{11}{15}.$$

e. For the average error probability from the point of view of the sender it follows that

$$P_e = \sum_{i=1}^{3} p(x_i)\left[1 - q(y_i/x_i)\right] = \frac{1}{3}(1 - 0.5) + \frac{1}{3}(1 - 0.3) + \frac{1}{3}(1 - 0)$$

$$= \frac{1}{6} + \frac{7}{30} + \frac{1}{3} = \frac{11}{15}.$$

The answers for d and e are identical. See Theorem 4.3.

f. The transition probability $q(y_3/x_3) = 0$, in other words the probability that a transmitted symbol x_3 goes over into a correct symbol y_3, is zero. On the other hand, the probability that a transmitted symbol x_2 goes over into an incorrect symbol y_3 is large, namely 0.9.

g. The amount of transported information follows from the relation

$$R = H(X) - H(X/Y) = H(Y) - H(Y/X) = 1.46 - 1.18$$
$$= 0.28 \text{ bit/symbol.}$$

The equivocation $H(X/Y)$ can also be calculated with the help of this relation, namely

$$H(X/Y) = H(X) - R.$$

For $H(X)$ we have

$$H(X) = \log 3 = 1.58 \text{ bits,}$$

so that

$$H(X/Y) = 1.58 - 0.28 = 1.30 \text{ bits/symbol.}$$

h. The Fano inequality reads

$$H(X/Y) \leq H(P_e) + P_e \log(n-1).$$

Calculating the first term on the right-hand side gives

$$H(P_e) = -P_e \log P_e - (1 - P_e) \log (1 - P_e)$$

$$= -\frac{11}{15} \log \frac{11}{15} - \frac{4}{15} \log \frac{4}{15} = 0.84.$$

The second term on the left-hand side gives

$$P_e \log (n-1) = \frac{11}{15} \log(3 - 1) = \frac{11}{15} = 0.73.$$

Summation of the last two equations yields

$$H(P_e) + P_e \log (n-1) = 0.84 + 0.73 = 1.57 > H(X/Y) = 1.30.$$

Thus the Fano inequality is satisfied.

4.7. *a.* The channel has the structure shown in Figure 4.10(a).
In this case we find for $H(Y/X)$ that

$$H(Y/X) = \sum_{i=1}^{2} \sum_{j=1}^{3} p(x_i) \, q(y_j/x_i) \log q(y_j/x_i)$$

$$= -\alpha \left[(1-p)\log(1-p) + p \log p\right] - (1-\alpha)\left[p \log p + (1-p)\log(1-p)\right]$$

$$= -p \log p - (1 - p) \log(1 - p).$$

b. To be able to calculate R we must first determine $H(Y)$. With

$$q(y_1) = \alpha(1 - p),$$

$$q(y_2) = \alpha p + (1 - \alpha)p = p,$$

$$q(y_3) = (1 - \alpha)(1 - p),$$

it follows that

$$H(Y) = -\alpha\,(1-p)\log \alpha(1-p) - p \log p - (1-\alpha)(1-p)\log(1-\alpha)(1-p\,).$$

After rewriting this becomes

$$H(Y) = -(1-p\,)\alpha\cdot\log \alpha - (1-p)(1-\alpha)\log(1-\alpha) - p \log p - (1-p)\log(1-p).$$

Hence for R we find

$$R = H(Y) - H(Y/X) = (1-p)\big\{-\alpha\log \alpha - (1-\alpha)\log(1-\alpha)\big\}.$$

c. The capacity follows from

$$C = \max_{\alpha} R.$$

Although we can determine the optimal value of α from $dR/d\alpha = 0$ we can find it directly from the expression for R by realising that the part between the curly brackets is really $H(\alpha, 1 - \alpha)$ and therefore maximally 1. Hence it directly follows that

$$C = 1 - p.$$

4.8. *a.* The transition matrix is as follows:

$$Q = \begin{bmatrix} 1 & 0 \\ p & 1-p \end{bmatrix}.$$

We therefore find for the probabilities $q(y_j)$

$$q(y_1) = 1 - \alpha + \alpha p; \qquad q(y_2) = \alpha(1 - p).$$

$H(Y)$ then becomes

$$H(Y) = -(1 - \alpha + \alpha p) \log (1 - \alpha + \alpha p) - \alpha (1-p) \log \alpha (1-p).$$

b. To calculate R we must first find $H(Y/X)$. We have

$$H(Y/X) = -\sum_{i=1}^{2}\sum_{j=1}^{2} p(x_1)\, q(y_j/x_i) \log q(y_j/x_i)$$

$$= -(1-\alpha)\cdot 1 \log 1 - \alpha p \log p - (1-\alpha)\cdot 0 \log 0 - \alpha(1-p)\log(1-p)$$

$$= -\alpha p \log p - \alpha(1-p)\log(1-p).$$

Hence

$$R = H(Y) - H(Y/X)$$

$$= -(1-\alpha + \alpha p) \log (1-\alpha + \alpha p) - \alpha (1-p) \log \alpha(1-p)$$

$$+ \alpha p \log p + \alpha (1-p) \log (1-p)$$

$$= -(1-\alpha + \alpha p) \log (1-\alpha + \alpha p) - \alpha (1-p) \log \alpha + \alpha p \log p.$$

c. To determine the capacity we differentiate the expression found for R with respect to α. This gives

$$\frac{dR}{d\alpha} = (1-p)\log(1-\alpha + \alpha p) - (1-\alpha + \alpha p) \frac{p-1}{1-\alpha + \alpha p} \log e$$

$$- (1-p)\log \alpha - \alpha(1-p) \frac{1}{\alpha} \log e + p \log p$$

$$= (1-p)\log(1-\alpha + \alpha p) - (1-p)\log \alpha + p \log p.$$

With $dR/d\alpha = 0$ it now successively follows for $\alpha = \alpha_0$ that

$$\log \frac{1 - \alpha_0 + \alpha_0 p}{\alpha_0} = -\frac{p}{1-p} \log p$$

and

$$1 - \alpha_0 + \alpha_0 p = \alpha_0 p^{p/(p-1)}$$

and

$$\alpha_0 = \frac{1}{1-p + p^{p/(p-1)}}.$$

We thus find for the capacity

$$C = -(1-\alpha_0 + \alpha_0 p)\log(1-\alpha_0 + \alpha_0 p) - \alpha_0(1-p)\log \alpha_0 + \alpha_0 p \log p$$

with

$$\alpha_0 = \frac{1}{1-p + p^{p/(p-1)}}.$$

After rewriting this gives

$$C = -\log \alpha_0 - \frac{p}{p-1} \log p = \log(1-p + p^{p/(p-1)}) - \frac{p}{p-1} \log p.$$

5

The continuous information source

5.1 Probability density functions

Before introducing measures of information for the continuous case some attention must first be paid to the concept of *probability density*. A stochastic variable may be obtained in the discrete case by mapping the outcomes of an experiment onto a number line and subsequently associating a probability with each number on this number line. For a continuous stochastic variable there are an infinite number of outcomes. In this case, events should in turn be defined to which a certain probability can be associated. By mapping the outcomes onto the number line via a given function, a continuous stochastic variable is defined with a range which is a (finite or infinite) interval on the number line.

Finding the probability of a certain outcome, that is of a certain value of the continuous stochastic variable, is not as easy as for a discrete stochastic variable. Since there are an infinite number of values, the probability of each value must be zero, while their sum must be equal to one. We may, however, speak of the probability that a value lies in a certain sub-interval of the number line. This corresponds with what we call an event. Thus we may say that for a voltage between -10 V and $+10$ V, the probability of exactly 6 V is zero, while the probability that the voltage lies between 5.9 V and 6.12 V may very well be 0.8 and is thus greater than zero. Events of the form ($a < x \leq b$) may be described with the help of cumulative distribution functions. The *cumulative distribution function* of a stochastic variable $\mathbf{x}$ is defined as the probability of the event $\{\mathbf{x} \leq x\}$:

$$F(x) = P(\mathbf{x} \leq x), \quad \text{for} - \infty < x < +\infty. \tag{5.1}$$

That means that $F(x)$ represents the probability that the stochastic variable takes on a value in the set $(-\infty, x]$.

155

The event $\{\mathbf{x} \leq x\}$ and its probability vary as x is varied. Thus $F(x)$ is a function of the variable x. For continuous variables this distribution function is a continuous function.

On the basis of the fundamentals of probability theory clearly the cumulative distribution function has the following properties:

i. $0 \leq F(x) \leq 1$,

ii. $\lim_{x \to \infty} F(x) = 1$,

iii. $\lim_{x \to -\infty} F(x) = 0$,

iv. $F(x)$ is non-decreasing in x, i.e. if $a < b$ then $F(a) \leq F(b)$.

In Figure 5.1 an example of a cumulative distribution function is given.

The probability density function (*pdf*) of $\mathbf{x}$, if it exists, is defined as the derivative of the cumulative distribution $F(x)$:

$$p(x) = \frac{dF(x)}{dx},\tag{5.2}$$

where $F(x)$ must be differentiable. An example of a probability density function is given in Figure 5.2.

A number of properties of the probability density function will be given. Since the cumulative distribution function is monotonic non-decreasing (property (iv)) it follows directly that

$$p(x) \geq 0.$$

And also

$$\int_{-\infty}^{x} p(u)\, du = F(x) - F(-\infty) = F(x).\tag{5.3}$$

Figure 5.1. Example of a distribution function.

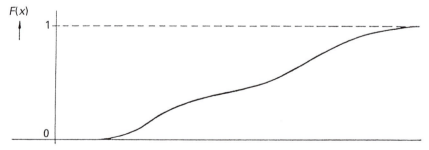

We thus see that the value of the distribution function corresponds to the shaded area in Figure 5.3. From this it can directly be seen that

$$\int_{-\infty}^{\infty} p(x)\, dx = F(\infty) - F(-\infty) = 1. \tag{5.4}$$

The area under the whole curve must therefore be equal to one. If we take a and b as the limits of integration, then (see Figure 5.4)

$$\int_{a}^{b} p(x)\, dx = F(b) - F(a) = P(\mathbf{x} \le b) - P(\mathbf{x} \le a) = P(a < \mathbf{x} \le b). \tag{5.5}$$

The probability that a continuous stochastic variable takes on a value between a and b can thus be found by integrating its probability density over (a,b). If $a = b$ it follows that $P(\mathbf{x} = a) = 0$ which is in agreement with the

Figure 5.2. Example of a probability density function.

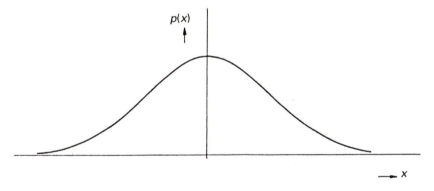

Figuur 5.3. Relationship between probability density and distribution function.

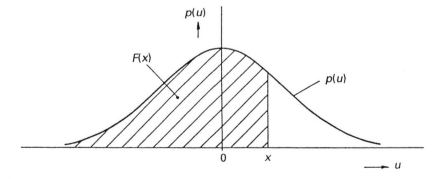

previously made statement that in the continuous case the probability of a specific value is equal to 0. The probability density of a stochastic variable fulfils much the same role as probabilities do for discrete stochastic variables. For a continuous stochastic variable we can only speak of a probability if the probability density is integrated over a certain interval (a,b). The probability density may also be regarded as a limit in this context. For a small value of Δx it follows that

$$P(x < \mathbf{x} \le x + \Delta x) \approx p(x)\, \Delta x,$$

since the area under the probability density curve over the interval $(x, x + \Delta x)$ can be approximated by a rectangle of width Δx. Hence

$$p(x) = \lim_{\Delta x \to 0} \frac{P(x < \mathbf{x} \le x + \Delta x)}{\Delta x}.$$

Notice that a probability density can be larger than one, which is also a difference compared with probabilities, which may at most be equal to one. The integral of $p(x)$ over a given interval, thus the probability that $\mathbf{x}$ takes on a value of x in this interval, must be less than or equal to one however.

A well-known continuous probability distribution is the uniform distribution. A continuous stochastic variable $\mathbf{x}$ has a uniform distribution, if the following holds for the probability density:

$$\left.\begin{aligned} p(x) &= \frac{1}{b-a} \quad &&\text{for } a \le x \le b, \\ &= 0 \quad &&\text{for } x < a,\, x > b. \end{aligned}\right\} \tag{5.6}$$

The distribution function of $\mathbf{x}$ is

Figure 5.4.

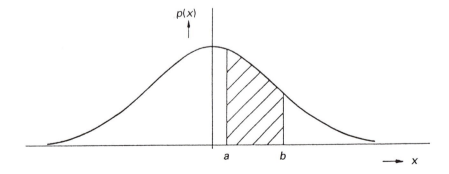

$$F(x) = 0 \qquad \text{for } x < a,$$

$$= \frac{x-a}{b-a} \qquad \text{for } a \le x \le b, \qquad \qquad (5.7)$$

$$= 1 \qquad \text{for } x > b.$$

Notice that for $b - a < 1$ the probability density is greater than one. The functions $p(x)$ and $F(x)$ are depicted in Figure 5.5.

Another well-known continuous probability distribution is the *normal or gaussian distribution*. This distribution is important because many phenomena have normally distributed stochastic variables, such as noise in communication systems and measurement errors made when observing systems and signals. A stochastic variable **x** has a normal or gaussian distribution if the probability density is given by

$$p(x) = \frac{1}{\sigma\sqrt{2\pi}} \exp\left\{ -\frac{(x-\mu)^2}{2\sigma^2} \right\} \qquad \text{for } -\infty < x < \infty, \quad (5.8)$$

where μ and σ are two parameters. For continuous distributions use is often made of the *mean* or the *expectation* of **x**, defined by

$$E(\mathbf{x}) = \int_{-\infty}^{\infty} x\, p(x)\, \mathrm{d}x, \qquad (5.9)$$

and of the *variance*. The variance is a measure for the variations of the values of **x** round its mean and is defined by

$$\mathrm{var}(\mathbf{x}) = E\left[(x - E(\mathbf{x}))^2\right] = \int_{-\infty}^{\infty} (x - E(\mathbf{x}))^2\, p(x)\, \mathrm{d}x. \qquad (5.10)$$

If we calculate $E(\mathbf{x})$ and $\mathrm{var}(\mathbf{x})$ for the gaussian distribution we find

Figure 5.5. Probability density and distribution functions for a uniform distribution.

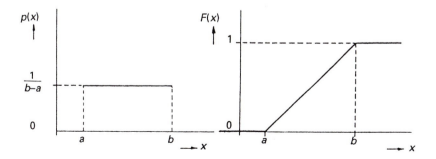

$$E(\mathbf{x}) = \mu \tag{5.11}$$

and

$$\mathrm{var}(\mathbf{x}) = \sigma^2. \tag{5.12}$$

From this it follows that the parameter μ represents the expectation of x and σ^2 the variance. For completeness it should be mentioned that the square root of the variance is called the *standard deviation* and in this case is thus equal to σ. The parameters μ and σ are characteristic for the form of $p(x)$. Hence the distribution is also often denoted by $N(\mu,\sigma)$. The parameter μ determines the point of symmetry of the graph of $p(x)$ and $\mu - \sigma$ and $\mu + \sigma$ the points of inflexion of this graph. (See Figure 5.6.)

The corresponding distribution function (see Figure 5.7) is

$$F(x) = \frac{1}{\sigma\sqrt{2\pi}} \int_{-\infty}^{x} \exp\left\{-\frac{(u-\mu)^2}{2\sigma^2}\right\} du. \tag{5.13}$$

Tables exist for a gaussian distribution function $F(x)$ with $\mu = 0$ and $\sigma = 1$. An arbitrary gaussian distribution $N(\mu,\sigma)$ can be normalized to a distribution

Figure 5.6. Gaussian distribution.

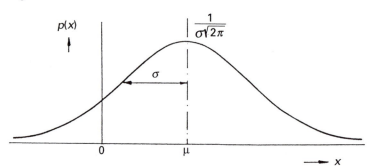

Figure 5.7. Distribution of a gaussian distribution.

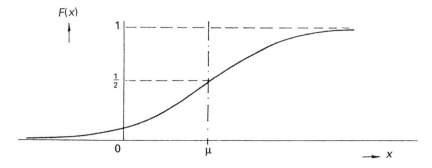

$N(0,1)$ by applying the transformation $\mathbf{y} = (\mathbf{x} - \mu)/\sigma$ to the stochastic variable $\mathbf{x}$. The distribution $N(0,1)$ is usually called the standard normal distribution. From the symmetry of the distribution it follows that $p(\mu + x) = p(\mu - x)$ and also that $F(\mu) = 1/2$, since then integration over half of the range of $\mathbf{x}$ will have taken place.

Besides considering one continuous stochastic variable we will also be dealing with combinations of continuous stochastic variables. Suppose we have two continuous variables $\mathbf{x}$ and $\mathbf{y}$ with probability densities $p(x)$ and $q(y)$. For each number pair (x,y) we now want to know $p(x,y)$. To this end we first consider the *joint cumulative distribution* for two continuous stochastic variables $\mathbf{x}$ and $\mathbf{y}$. This is defined as

$$F(x,y) = P(\mathbf{x} \le x, \mathbf{y} \le y). \tag{5.14}$$

The *joint* or *two-dimensional probability density* $p(x,y)$ is now defined as the partial derivative with respect to x and y.

$$p(x,y) = \frac{\partial^2 F(x,y)}{\partial x\, \partial y}. \tag{5.15}$$

An example of a two-dimensional probability density is given in Figure 5.8. It is still true that

Figure 5.8. Example of a two-dimensional probability density function.

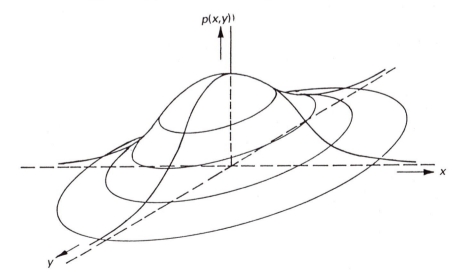

$$\int\limits_{-\infty}^{\infty}\int\limits_{-\infty}^{\infty} p(x,y)\,dx\,dy = 1, \tag{5.16}$$

which means that the volume under the joint probability density function must be equal to one. Just as there was a relation between the marginal and joint probability in the discrete case we now have

$$p(x) = \int\limits_{-\infty}^{\infty} p(x,y)\,dy, \tag{5.17}$$

$$q(y) = \int\limits_{-\infty}^{\infty} p(x,y)\,dx. \tag{5.18}$$

As an example of a two-dimensional probability density function consider the two-dimensional gaussian distribution. This is expressed as

$$p(x_1,x_2) = \frac{1}{2\pi\sigma_1\sigma_2\sqrt{1-\rho^2}}\cdot\exp\left\{-\frac{1}{2(1-\rho^2)}\left[\frac{(x_1-\mu_1)^2}{\sigma_1^2}\right.\right.$$
$$\left.\left. -\frac{2\rho(x_1-\mu_1)\cdot(x_2-\mu_2)}{\sigma_1\sigma_2} + \frac{(x_2-\mu_2)^2}{\sigma_2^2}\right]\right\}, \tag{5.19}$$

with parameters μ_1, μ_2, σ_1, σ_2 and ρ. Here, ρ is the *correlation coefficient*, where $-1 < \rho < 1$ and on which more will be said later. The probability density is depicted in Figure 5.8 for the case $\mu_1 = \mu_2 = 0$, $\sigma_1 = \sigma_2$, and $\rho = 0$. The corresponding marginal probability densities $p(x)$ and $p(y)$, which are obtained by applying equations (5.17) and (5.18), are also distributed normally according to $N(\mu_1,\sigma_1)$ and $N(\mu_2,\sigma_2)$ respectively.

For continuous stochastic variables which are dependent on one another, one can make use of *conditional probability densities*. The use of conditional probability densities is very important for problems in electronics and information technology. This is because noise signals have a continuous range, which means that problems such as separating information carrying signals and noise signals and interpreting measurements disturbed by noise are by nature formulated in terms of a conditional probability density (or probability), where the condition is formed by the measured results. The definition of the conditional probability density function is, in analogy with the discrete case, formulated in terms of the marginal and joint probability density. We have

$$p(x/y) = \frac{p(x,y)}{q(y)}. \tag{5.20}$$

The relationship between the marginal probability density function $p(x)$ and the conditional probability density $p(x/y)$ is given by

$$p(x) = \int_{-\infty}^{\infty} p(x,y)\,\mathrm{d}y = \int_{-\infty}^{\infty} q(y)\,p(x/y)\,\mathrm{d}y. \tag{5.21}$$

Bayes' formula can also be given for probability densities as

$$p(x/y) = \frac{p(x)\,p(y/x)}{q(y)}, \tag{5.22}$$

or also

$$p(x/y) = \frac{p(x)\,p(y/x)}{\int_{-\infty}^{\infty} p(x)\,p(y/x)\,\mathrm{d}x}. \tag{5.23}$$

For completeness it should be added that the last three expressions also hold if x is a continuous variable and y a discrete one, in which case these expressions then take on the following form:

$$p(x) = \sum_{i} q(y_i)\,p(x/y_i),$$

$$p(x/y_i) = \frac{p(x)\,q(y_i/x)}{q(y_i)},$$

$$p(x/y_i) = \frac{p(x)\,q(y_i/x)}{\int_{-\infty}^{\infty} p(x)\,q(y_i/x)\,\mathrm{d}x}.$$

These relations are used in information theory and especially in statistical detection theory. In analogy with the discrete case, statistical independence of two continuous stochastic variables x and y can be defined as

$$p(x,y) = p(x)\,q(y). \tag{5.24}$$

In that case the following are also true:

$$q(y/x) = q(y),$$

$$p(x/y) = p(x).$$

Through the use of conditional probabilites and conditional probability density functions we can indicate that dependence exists between two continuous stochastic variables. The *covariance* or *correlation* is also often

used instead. Let **x** and **y** be two continuous stochastic variables, then the
covariance is defined by

$$\text{cov}(\mathbf{x},\mathbf{y}) = E\big[(x - E(\mathbf{x}))\cdot(y - E(\mathbf{y}))\big]$$

$$= \int\limits_{-\infty}^{\infty} \int\limits_{-\infty}^{\infty} (x - E(\mathbf{x}))\,(y - E(\mathbf{y}))\,p(x,y)\,dxdy. \tag{5.25}$$

Notice that the covariance of the stochastic variable **x** with itself is

$$\text{cov}(\mathbf{x},\mathbf{x}) = E\big[(x - E(\mathbf{x}))^2\big] = \text{var}(\mathbf{x}). \tag{5.26}$$

If the covariance is normalised with respect to the variances of **x** and **y** then
the *correlation coefficient* ρ is obtained, which we previously encountered in
the expression for the two-dimensional gaussian distribution. The
correlation coefficient is given by

$$\rho = \frac{\text{cov}(\mathbf{x},\mathbf{y})}{\sqrt{\text{var}(\mathbf{x})\,\text{var}(\mathbf{y})}}. \tag{5.27}$$

It can be shown that $|\rho| \le 1$.
The *correlation* between **x** and **y** is defined by

$$R(\mathbf{x},\mathbf{y}) = E(\mathbf{x}\cdot\mathbf{y}) = \int\limits_{-\infty}^{\infty} \int\limits_{-\infty}^{\infty} x\,y\,p(x,y)\,dxdy. \tag{5.28}$$

A relation between the covariance and the correlation can be derived with
the help of equation (5.25):

$$\text{cov}(\mathbf{x},\mathbf{y}) = \int\limits_{-\infty}^{\infty} \int\limits_{-\infty}^{\infty} \{xy - x\,E(\mathbf{y}) - y\,E(\mathbf{x}) + E(\mathbf{x})E(\mathbf{y})\}\,p(x,y)\,dxdy$$

$$= \int\limits_{-\infty}^{\infty} \int\limits_{-\infty}^{\infty} xy\,p(x,y)\,dxdy - E(\mathbf{x})\,E(\mathbf{y}) - E(\mathbf{y})E(\mathbf{x}) + E(\mathbf{x})E(\mathbf{y})$$

$$= R(\mathbf{x},\mathbf{y}) - E(\mathbf{x})E(\mathbf{y}). \tag{5.29}$$

5.2 Stochastic signals

Signals of which the amplitude at an arbitrary point in time is time
dependent are called *stochastic signals*. Formally we define a stochastic
signal as $\{\mathbf{x}(t),\, t \in T\}$, where T can be a certain time interval, but also a set
of points of time (the beginning of each hour for example). With $\mathbf{x}(t)$ we
mean the value of the signal at the point of time t.

At a given point of time t_0, $\mathbf{x}(t_0)$ is a stochastic variable that determines which values the signal can take at time t_0 and of which it is assumed to be known how probable it is that these values do indeed occur. Because the signal $\mathbf{x}(t)$ is given for all t out of a set T this in principle determines which forms the signal $\mathbf{x}(t)$ can take on as a whole and how probable it is that they occur. One can thus regard a stochastic signal $\mathbf{x}(t)$ for each point of time $t \in T$ as a stochastic variable. On the other hand one can also determine what the signal looks like each time, for all the points of time in the set T. The realized values of the signal are then regarded as a whole. We then speak of a *realization* of the stochastic signal, denoted by $x(t)$. A realization of a stochastic signal can be regarded as a result of an experiment: evoke the stochastic signal $\{\mathbf{x}(t), t \in T\}$ once and observe the result (that is, the consequent signal values $[x(t), t \in T]$). The set of all possible realizations is often denoted as the *ensemble*.

With analogue signals, the signal value at each point of time is a continuous quantity, so that the stochastic process $\{\mathbf{x}(t), t \in T\}$ is a continuous stochastic variable for a given point of time t with a certain probability density $p(x)$. The signal is also a continuous phenomenon as a function of time, which means that the set T is a certain time interval. This is often the measurement time or observation time, but it may also be the intrinsic time that the signal occurs.

In general there are four types of stochastic signals which are summarized in Figure 5.9.

In the foregoing it has implicitly been assumed that the signals are one-dimensional signals as a function of time. We can, however, also consider two- or even more-dimensional signals, television images or multi-channel registrations of seismic signals for example. One can also regard the intensity of a television image as a stochastic process where the (x,y)-co-ordinates of the pixels on the screen fulfil a similar role to the time parameter.

Although an analogue signal has a continuous character with regard to both the signal values and time as a parameter, this description is too complex to be functional. In practice we often limit ourselves to so-called discrete time signals which are only known at certain points of time. Although this seems to be a limitation, this often turns out not to be the case, namely in those cases where the stochastic signal has a limited bandwidth.

This relationship between bandwidth and the time interval between consecutive points of time at which a signal should be determined is given by the *sampling theorem*.

Theorem 5.1 (Sampling theorem in time domain)
If a signal $x(t)$ has a bandwith W Hz, i.e. W cycles per second, then the signal is completely determined by giving its ordinates at a series of points spaced $1/2W$ seconds apart, the series extending throughout the time domain.
The continous signal $x(t)$ can be reconstructed from the samples $x(k/2W)$ as follows:

Figure 5.9. Types of stochastic signals.

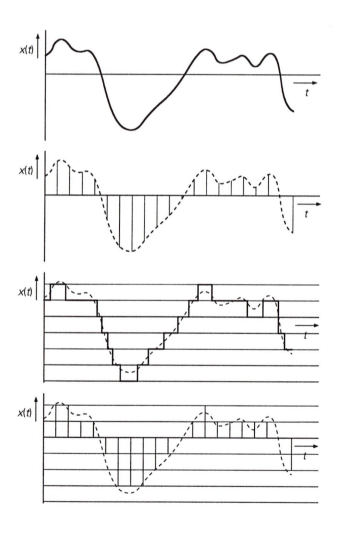

$$x(t) = \sum_{k=1}^{2WT} x\left(\frac{k}{2W}\right) \frac{\sin 2\pi W\left(t - \frac{k}{2W}\right)}{2\pi W\left(t - \frac{k}{2W}\right)}. \tag{5.30}$$

□

The proof of the theorem will not be given here. The function $\sin 2\pi Wt/2\pi Wt$, also denoted by sinc $2\pi Wt$, has the properties that its value is 1 at $t = 0$ and 0 at $t = k/2W$, i.e. all the sample points excepting $t = 0$ (see Figure 5.10).

The original signal $x(t)$ can be reconstructed from the values at the sample point, by using the sinc-function at each sample point multiplied by $x(k/2W)$. The superposition of all these sinc-functions gives $x(t)$. See Figure 5.11.

Thus, for band limited signals a time-discrete representation is sufficient, even if they are originally time-continuous. With regard to the samples of a time-discrete stochastic signal, denoted by $x(k/2W)$, it is often arranged for simplicity that $2W = 1$, so that we can use the notation $x(k)$, $k = 0,1,2,...$. In general, if the duration of the signal is T this will result in $2WT$ samples of the signal being available. Considering the time-discrete signal makes analysis easier, without the results losing generality.

Returning to stochastic signals $\{x(t), t \in T\}$ we can represent a signal, after applying the sampling theorem, as a series of (time-discrete) samples, where the value of each sample can be regarded as a stochastic variable. To determine the complete stochastic signal one must determine the joint probability density $p(\mathbf{x}) = p(x(t_1),x(t_2),...,x(t_N))$ for the total set of N sampling instants. Each "value" of $\mathbf{x}$ is then in fact a realization of the

Figure 5.10. Sinc-function.

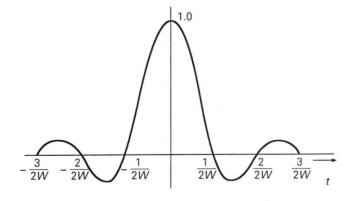

stochastic process $\{\mathbf{x}(t), t \in T\}$ determined at the sampling instants $t_1, t_2, \ldots, t_N$.

It is often unneccessary or even imposible to use the joint probability density of all samples. If we limit ourselves to a first-order description, that is with $N = 1$, then we can consider the signal characteristics per sample. Since dependence between samples plays a large role in practice, a second-order description is usually desirable or required, that is with $N = 2$. This is usually also sufficient. A second-order stochastic signal may be determined with $\{\mathbf{x}(t_i,t_j), t_i, t_j \in T\}$ or in other words through two samples with joint probability density $p(x(t_i), x(t_j))$.

One can in principle analyse the signal for a given probability density $p(x(t_i), x(t_j))$. However, quantities such as the expectation, the correlation or the covariance are often examined and not the probability density itself.

The *autocorrelation function* is defined as

$$R_{xx}(t_i,t_j) = E\{\mathbf{x}(t_i)\cdot\mathbf{x}(t_j)\} = \int\limits_{-\infty}^{\infty} \int\limits_{-\infty}^{\infty} x(t_i)\, x(t_j)\, p(x(t_i),x(t_j))\, \mathrm{d}x(t_i)\mathrm{d}x(t_j),$$

(5.31)

and thus depends on t_i and t_j.

If we now use the sampling instants for t_i and t_j then we can also determine the values of the autocorrelation function with the help of a matrix, the so-called *autocorrelation matrix* $\mathbf{R}_{xx}$:

$$\mathbf{R}_{xx} = \begin{bmatrix} R_{11} & \cdots & R_{1N} \\ \vdots & & \vdots \\ R_{N1} & \cdots & R_{NN} \end{bmatrix},$$

(5.32)

where $R_{ij} = R_{xx}(t_i,t_j)$.

This definition is still very general since the correlation between two samples $x(t_i)$ and $x(t_j)$ may be different for different pairs of points of time t_i

Figure 5.11. Illustration of sampling theorem.

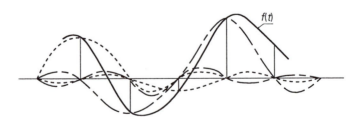

and t_j. Such signals are called *non-stationary*. One can often assume, however, that the signals are *stationary* in that the correlation depends only on the difference in time $\tau = t_i - t_j$ and not on the absolute time instants t_i and t_j. We then speak of a *weak stationary signal*, for which we have that

$$R_{xx}(t_i,t_j) = R_{xx}(t_i - t_j) = R_{xx}(\tau) = E\{\mathbf{x}(t_i)\,\mathbf{x}(t_i - \tau)\}. \tag{5.33}$$

For $\tau = 0$ we find $R_{xx(0)} = E\{x(t_i)^2\}$ which is the *average power* per sample, also denoted by P_x.

In the case of a weak stationary signal it is possible that the probability density $p(x(t_i),x(t_j))$ of a *weak stationary signal* does still depend on the absolute points of time t_i and t_j. The term *strictly stationary* is therefore also used to indicate that the probability density $p(x(t_i),x(t_j))$ itself is time invariant, i.e. depends only on $t_i - t_j$. A strictly stationary signal is also a weak stationary signal, but the converse is thus not *per se* true.

Besides the autocorrelation, use is also often made of the *autocovariance*. This is defined as

$$K_{xx}(t_i,t_j)$$
$$= \int\limits_{-\infty}^{\infty} \int\limits_{-\infty}^{\infty} (x(t_i) - E\{\mathbf{x}(t_i)\}) \cdot (x(t_j) - E\{\mathbf{x}(t_j)\})\, p(x(t_i),x(t_j)) \mathrm{d}x(t_i)\mathrm{d}x(t_j),$$

$$\tag{5.34}$$

The difference from the correlation is purely that the expectation of $x(t_i)$ and $x(t_j)$ is now subtracted from $x(t_i)$ and $x(t_j)$. For many applications these expected values are zero, so that the correlation and covariance are then equal to each other. The *autocovariance matrix* for N samples is

$$\mathbf{K}_{xx} = \begin{bmatrix} K_{11} & \dots & K_{1N} \\ \vdots & & \vdots \\ K_{N1} & \dots & K_{NN} \end{bmatrix}, \tag{5.35}$$

where $K_{ij} = K_{xx}(t_i,t_j)$. It is simple to see that the autocorrelation matrix and the autocovariance matrix are symmetric, that is $R_{ij} = R_{ji}$ and $K_{ij} = K_{ji}$.

An important class of signals, for which it is sufficient that the signal is weak stationary, is formed by gaussian stochastic signals. For these signals the autocovariance matrix is a sufficient description of the signal.

A stochastic signal $\{\mathbf{x}(t), t \in T\}$ is called a *gaussian signal* if all of its N-dimensional probability densities $p(x(t_1),\dots,x(t_N))$ for $N = 1,2,\dots$ are N-dimensional gaussian probability densities.

The gaussian probability density for $N = 1$, thus for one sample, is given in equation (5.8). For $N = 2$ the two-dimensional gaussian distribution is given in equation (5.19).

In the general case of an N-dimensional gaussian distribution it is simple to use a vector notation. The expression for $\tilde{x}(x(t_1),\ldots,x(t_N))$ is then

$$p(\tilde{x}) = \frac{1}{(2\pi)^{N/2}|\mathbf{K}_{xx}|^{1/2}} \exp\{-\tfrac{1}{2}(\tilde{x} - \tilde{\mu})\,\mathbf{K}_{xx}^{-1}(\tilde{x} - \tilde{\mu})^T\}, \qquad (5.36)$$

where $\tilde{\mu}$ is the vector of the mean values, and $|\mathbf{K}_{xx}|$ the determinant of the autocovariance matrix $\mathbf{K}_{xx}$.
The gaussian signal is the only signal that is determined completely by its mean and its autocovariance matrix.

If the process is stationary, the covariance K_{ij} only depends on the time difference $|t_i - t_j|$. This means for example that $K_{ij} = K_1$ for all points of time t_i and t_j such that $t_j = t_i + 1$ or $t_j = t_i - 1$. The autocovariance matrix then has the form

$$\mathbf{K}_{xx} = \begin{bmatrix} K_0 & K_1 & \cdots & \cdots & K_{N-1} \\ K_1 & K_0 & \cdots & \cdots & \cdots \\ \cdots & \cdots & \cdots & \cdots & \cdots \\ \cdots & \cdots & \cdots & \cdots & K_1 \\ K_{N-1} & \cdots & \cdots & K_1 & K_0 \end{bmatrix}, \qquad (5.37)$$

and thus exhibits a very specific structure. A matrix of this form is called a *Toeplitz matrix*. The first row of this matrix gives the sampled values of the autocovariance function of the gaussian process.

A gaussian signal has a number of properties which make its use attractive. Possibly the most important property is that a linear manipulation which is carried out on a gaussian signal leads to a signal which is also gaussian. Because of this, the gaussian process plays a role with the analysis of stochastic signals that is comparable with the role of linear systems in systems theory.

The description of stochastic signals has thus far remained limited to the time domain. However, one can also use the frequency domain here. Then the *power density spectrum* is very important, which is a measure for the amount of power per bandwidth of the stochastic signal.
This *power density spectrum* $S_{xx}(\omega)$, often simply called the spectrum, can be found from the autocorrelation function with the help of a Fourier transformation. It is defined as follows:

$$S_{xx}(\omega) = \int_{-\infty}^{\infty} R_{xx}(\tau)\, e^{-j\omega\tau} d\tau, \tag{5.38}$$

$$R_{xx}(\tau) = \frac{1}{2\pi} \int_{-\infty}^{\infty} S_{xx}(\omega)\, e^{+j\omega\tau} d\omega. \tag{5.39}$$

For the case $\tau = 0$ we find

$$R_{xx}(0) = \frac{1}{2\pi} \int_{-\infty}^{\infty} S_{xx}(\omega)\, d\omega = P_x. \tag{5.40}$$

This means that the average power of a signal can be found from the autocorrelation function by setting $\tau = 0$ as seen earlier, but also by integrating the power density spectrum over the whole frequency range.

5.3 The continuous information measure

The continuous information measure can be derived on the basis of the discrete one as follows. As seen above, a continuous probability density function can be approximated by means of probability densities that are constant on intervals of length Δx. Assume that p_i is the value of the constant probability p_i at the interval i. To guarantee that $\sum p_i = 1$, set $p_i = p(x_i)\Delta x$, where x_i is a point in the interval i such that $p(x_i)\Delta x$ is equal to the area under the continuous probability density function $p(x)$ in interval i. Now it is the case that

$$H(X) = -\sum_{i=1}^{n} p_i \log p_i$$

$$= -\sum_{i=1}^{n} p(x_i)\Delta x \log p(x_i)\Delta x$$

$$= -\sum_{i=1}^{n} p(x_i)\Delta x \log p(x_i) - \log \Delta x. \tag{5.41}$$

Taking the limit for $\Delta x \to 0$ we find

$$\lim_{\Delta x \to 0} H(X) = -\int_{-\infty}^{\infty} p(x) \log p(x) - \lim_{\Delta x \to 0} \log \Delta x. \tag{5.42}$$

Clearly, the second term will become infinite. Thus the amount of information of a continuous random variable is always infinite. In fact this result is not surprising. Interpreting the information measure as giving the average number of yes/no answers necessary to resolve the uncertainty, this number will be infinite in the continuous case.

However, this is just a theoretical point of view. In practice there will be some measurement uncertainty.

This also makes Δx finite. In general Δx is chosen to be equal to the unit, by which the second term equals zero. This leads to the following definition.

Definition 5.1
For the continuous stochastic variable **x** with probability density function $p(x)$ the amount of information is equal to

$$H(X) = - \int_{-\infty}^{\infty} p(x) \log p(x) \, dx. \qquad (5.43)$$

□

Clearly, the definition of the continuous infomation measure is based on analogy with the discrete one, rather than on a sound mathematical derivation.

As a consequence the $H(X)$ defined in this manner for a continuous stochastic variable can become negative, which is in contrast to the discrete case.

It is also important to know what probability density leads to a maximum amount of information for a continuous stochastic variable. The derivation differs from that given for the discrete case, however. This is due to the fact that extra restrictions usually have to be imposed for continuous stochastic variables. These restrictions may for example be a bounded amplitude or a constant power (variance). The nature of these restrictions thus jointly determines the nature of the probability density which leads to a maximum amount of information. Two cases will be considered, namely amplitude bounding, and bounding of the power (or variance). In the following theorem, the probability density which leads to a maximum amount of information will be determined if the range is limited to between $-A$ and $+A$.

Theorem 5.2
For a signal which is bounded in amplitude within the range $(-A, +A)$, the amount of information $H(X)$ is maximum if and only if

$$p(x) = \frac{1}{2A}.$$

The maximum value is given by

$$H(X) = \log 2A. \tag{5.44}$$

Proof

To solve this problem use will be made of a method from the calculus of variations. The task is to determine the probability density $p(x)$ for which

$$H(X) = -\int_{-\infty}^{\infty} p(x) \log p(x) \, dx,$$

is maximum, where due to the bounded amplitude $p(x)$ must satisfy

$$\int_{-A}^{A} p(x) \, dx = 1.$$

To this end the function

$$G(x) = -p(x) \log p(x) + \alpha p(x)$$

is differentiated with respect to $p(x)$ and subsequently set equal to zero. This gives

$$-\log p(x) - \log e + \alpha = 0$$

or

$$\ln p(x) = \frac{\alpha}{\log e} - 1 = k, \quad \text{so that} \quad p(x) = e^k.$$

We have

$$\int_{-A}^{A} p(x) \, dx = 1,$$

so that substitution $p(x)$ gives

$$\int_{-A}^{A} e^k \, dx = 1 \quad \Rightarrow \quad \left[e^k x \right]_{-A}^{A} = e^k 2A = 1,$$

which yields

$$p(x) = \frac{1}{2A}.$$

Substitution of $p(x) = \frac{1}{2A}$ in $H(X)$ yields

$$H(X) = -\int_{-A}^{A} \frac{1}{2A} \log\left(\frac{1}{2A}\right) dx = \log 2A.$$ $\square$

Evidently in the case of amplitude bounding, the uniform probability density delivers a maximum amount of information proportional to the maximum amplitude. This is partly in agreement with the case of a discrete stochastic variable, which has a maximum amount of information if it has a uniform probability distribution.

Another important case is when the power of a signal is bounded, which comes down to fixing the variance of the samples.

Theorem 5.3
For a signal with constant power σ^2,

$$\sigma^2 = \int_{-\infty}^{\infty} x^2 p(x)\, dx,$$

$H(X)$ is maximum if and only if

$$p(x) = \frac{1}{\sigma\sqrt{2\pi}} \exp\left\{-\frac{x^2}{2\sigma^2}\right\}.$$

The corresponding maximum amount of information is

$$H(X) = \log(\sigma\sqrt{2\pi\, e}).$$ (5.45)

Proof
We must now determine the probability density $p(x)$, such that

$$H(X) = -\int_{-\infty}^{\infty} p(x) \log p(x)\, dx$$

is maximum, where the following restrictions must be met:

$$\int_{-\infty}^{\infty} p(x)\, dx = 1,$$

and

$$\int_{-\infty}^{\infty} x^2 p(x)\, dx = \sigma^2,$$

where σ^2 is thus assumed to be constant. We now form the function

$$G(x) = -p(x)\log p(x) + \alpha_1 p(x) + \alpha_2 x^2 p(x),$$

and set the derivative of $G(x)$ with respect to $p(x)$ equal to zero. This gives

$$-\log p(x) - \log e + \alpha_1 + \alpha_2 x^2 = 0$$

or after dividing by $-\log_2 e$:

$$\ln p(x) + 1 - \lambda_1 - \lambda_2 \cdot x^2 = 0$$

with

$$\lambda_1 = \frac{\alpha_1}{\log e} \quad \text{and} \quad \lambda_2 = \frac{\alpha_2}{\log e}.$$

This yields the following solution:

$$p(x) = e^{\lambda_1 - 1} e^{\lambda_2 x^2}.$$

The parameters λ_1 and λ_2 are eliminated by substituting $p(x)$ in both restrictions:

$$\int_{-\infty}^{\infty} p(x)\,dx = \int_{-\infty}^{\infty} e^{\lambda_1 - 1} e^{\lambda_2 x^2}\,dx = 1.$$

This gives

$$e^{\lambda_1 - 1} = \sqrt{-\frac{\lambda_2}{\pi}}.$$

Furthermore,

$$\int_{-\infty}^{\infty} x^2 p(x)\,dx = \int_{-\infty}^{\infty} x^2 e^{\lambda_2 x^2} \sqrt{-\frac{\lambda_2}{\pi}}\,dx = \sigma^2,$$

from which

$$\lambda_2 = -\frac{1}{2\sigma^2},$$

and therefore

$$e^{\lambda_1 - 1} = \frac{1}{\sigma\sqrt{2\pi}}.$$

This yields finally

$$p(x) = \frac{1}{\sigma\sqrt{2\pi}} \exp\left\{-\frac{x^2}{2\sigma^2}\right\}.$$

The corresponding amount of information is

$$
\begin{aligned}
H(X) &= -\int_{-\infty}^{\infty} \frac{1}{\sigma\sqrt{2\pi}} \exp\left\{-\frac{x^2}{2\sigma^2}\right\} \log \frac{1}{\sigma\sqrt{2\pi}} \exp\left\{-\frac{x^2}{2\sigma^2}\right\} dx \\
&= \log \sigma\sqrt{2\pi} \int_{-\infty}^{\infty} \frac{1}{\sigma\sqrt{2\pi}} \exp\left\{-\frac{x^2}{2\sigma^2}\right\} dx \\
&\quad + \int_{-\infty}^{\infty} \frac{x^2 \log e}{2\sigma^2} \cdot \frac{1}{\sigma\sqrt{2\pi}} \exp\left\{-\frac{x^2}{2\sigma^2}\right\} dx \\
&= \log \sigma\sqrt{2\pi} + \frac{\log e}{2\sigma^2} \mathrm{var}(\mathbf{x}) = \log \sigma\sqrt{2\pi} + \frac{1}{2}\log e \\
&= \log \sigma\sqrt{2\pi e}. \qquad\qquad \square
\end{aligned}
$$

We thus find that the normal distribution yields a maximum amount of information for a constant power σ^2. $H(X)$ is proportional to the logarithm of the standard deviation σ. This is an important result since both the power and the normal distribution are frequently used in technical applications.

5.4 Information measures and sources with memory

Regarding the discrete case, we have defined the conditional, joint and mutual information measures in addition to the marginal information measure. This can be done for the continuous case as well. The conditional, joint and mutual information measures are briefly introduced for the continuous case below. Notice that the interrelations between the various measures by and large conform with those for the discrete case.

In the case where there are two stochastic variables $\mathbf{x}$ and $\mathbf{y}$ with joint probability density $p(x,y)$ the *joint amount of information* is defined as

$$H(X,Y) = -\int_{-\infty}^{\infty} \int_{-\infty}^{\infty} p(x,y) \log p(x,y)\, dx dy. \tag{5.46}$$

The conditional amount of information can be found as follows. The joint probability density for two stochastic variables $\mathbf{x}$ and $\mathbf{y}$ can be written as

$$p(x,y) = p(x)\cdot q(y/x) = q(y)\cdot p(x/y). \tag{5.47}$$

We now define the *conditional amount of information* of x given y as

$$H(X/Y) = - \int\limits_{-\infty}^{\infty} \int\limits_{-\infty}^{\infty} p(x,y) \log p(x/y) \, dxdy. \tag{5.48}$$

Similarly:

$$H(Y/X) = - \int\limits_{-\infty}^{\infty} \int\limits_{-\infty}^{\infty} p(x,y) \log q(y/x) \, dxdy. \tag{5.49}$$

These definitions thus conform strongly with the previously given definitions for the discrete case. Just as in the discrete case we can derive that

$$H(X/Y) \le H(X), \tag{5.50}$$

$$H(Y/X) \le H(Y), \tag{5.51}$$

with equality if **x** and **y** are statistically independent. We also have

$$H(X,Y) = H(X) + H(Y/X) = H(Y) + H(X/Y). \tag{5.52}$$

The proofs will not be given here but they proceed identically to the proofs of the discrete case. On the basis of the foregoing it is also the case that

$$H(X,Y) \le H(X) + H(Y), \tag{5.53}$$

with equality if **x** and **y** are statistically independent. We cannot, however, say that $H(Y/X) \ge 0$. If we define the mutual amount of information as

$$I(X;Y) = H(X) + H(Y) - H(X,Y), \tag{5.54}$$

then with the help of the previously given definitions we find that $I(X;Y)$ is equal to

$$I(X;Y) = \int\limits_{-\infty}^{\infty} \int\limits_{-\infty}^{\infty} p(x,y) \log \frac{p(x,y)}{p(x) \cdot q(y)} \, dxdy. \tag{5.55}$$

Example 5.1
Let the joint probability density function $p(x,y)$ be given by

$$p(x,y) = \tfrac{1}{4} \text{ for } 0 \le x \le 2 \text{ and } 0 \le y \le 4 - 2x,$$

$$= 0 \text{ otherwise.}$$

For the marginal probability density functions, it follows that

$$p(x) = \int_0^{4-2x} p(x,y)\, dy = \int_0^{4-2x} \frac{1}{4}\, dy = \frac{1}{4} y \Big|_0^{4-2x} = 1 - \frac{1}{2} x \quad \text{for } 0 \le x \le 2,$$

and

$$q(y) = \int_0^{2-\frac{1}{2}y} p(x,y)\, dy = \int_0^{2-\frac{1}{2}y} \frac{1}{4}\, dx = \frac{1}{4} x \Big|_0^{2-\frac{1}{2}y} = \frac{1}{2} - \frac{1}{8} y \quad \text{for } 0 \le y \le 4.$$

The amount of information can now be calculated directly, by taking into account that in general

$$\int x^n \ln x\, dx = \frac{x^{n+1}}{n+1} \left\{ \ln x - \frac{1}{n+1} \right\}.$$

For $H(X)$ we obtain

$$
\begin{aligned}
H(X) \;=\; & -\int_0^2 \left(1 - \frac{1}{2} x\right) \log \left(1 - \frac{1}{2} x\right) dx \\[2mm]
=\; & 2 \int_1^0 z \log z\, dz = 2 \log e \int_1^0 z \ln z\, dz \\[2mm]
=\; & 2 \log e\Big[\frac{1}{2} z^2 \ln z - \frac{1}{4} z^2 \Big]_1^0 = \log \sqrt{e} \approx 0.72.
\end{aligned}
$$

Similarly it can be proven that

Figure 5.12. Definition area of $p(x,y)$.

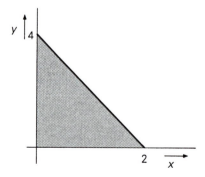

$$H(Y) = -\int\limits_0^4 (\tfrac{1}{2} - \tfrac{1}{8}y) \log (\tfrac{1}{2} - \tfrac{1}{8}y) \, dy = \log 2\sqrt{e} \approx 1.72$$

The conditional information measures can be calculated by determining the conditional probability density functions.

$$p(x/y) = \frac{p(x,y)}{q(y)} = \frac{\tfrac{1}{4}}{\tfrac{1}{2} - \tfrac{1}{8}y} = \frac{2}{4 - y}, \qquad 0 \le x \le 2 - \tfrac{1}{2}y,$$

$$p(y/x) = \frac{p(x,y)}{p(x)} = \frac{\tfrac{1}{4}}{1 - \tfrac{1}{2}x} = \frac{1}{4 - 2x}, \qquad 0 \le y \le 4 - 2x.$$

By substitution into the conditional information measure it follows that

$$H(X/Y) = \log \frac{2}{\sqrt{e}} \approx 0.28,$$

and

$$H(Y/X) = \log \frac{4}{\sqrt{e}} \approx 1.28.$$

These results can also be obtained indirectly via $H(X,Y)$.

$$H(X,Y) = -\int\limits_0^2 \int\limits_0^{4-2x} p(x,y) \log p(x,y)$$

$$= -\int\limits_0^2 \int\limits_0^{4-2x} \tfrac{1}{4} \log \tfrac{1}{4} \, dy dx$$

$$= \tfrac{1}{2}\int\limits_0^2 (4 - 2x) dx = \tfrac{1}{2} \left[4x - x^2 \right]_0^2 = 2.$$

From $H(X,Y)$ and $H(Y)$ it follows that

$$H(X/Y) = H(X,Y) - H(Y) = 2 - \log 2\sqrt{e} = \log \frac{2}{\sqrt{e}},$$

and

$$H(Y/X) = H(X,Y) - H(X) = 2 - \log \sqrt{e} = \log \frac{4}{\sqrt{e}}. \qquad \triangle$$

The continuous joint and conditional information measures play a role in the description of continuous information sources with memory.

The consecutive samples of a signal generated by such a source are generally dependent. For the discrete information source this dependence is expressed by the transition probabilities to the various symbols or states in a Markov chain. In the case of continuous information this can be done in terms of conditional probability density functions or the autocorrelation matrix $\mathbf{R}_{xx}$.

Suppose, in accordance with the sampling theorem, that we have $N = 2wT$ samples: $x(t_1), \dots, x(t_N)$. Due to the memory effect of the source these samples are not statistically independent. Because of this we should use the joint probability density functions in order to calculate the amount of information. This yields

$$H(\widetilde{X}) = -\int_{-\infty}^{\infty} p(\widetilde{x}) \log p(\widetilde{x}) \, d\widetilde{x}$$

$$= -\int_{-\infty}^{\infty} \dots \int_{-\infty}^{\infty} p(x_1,\dots,x_N) \log p(x_1,\dots,x_N) dx_1\dots dx_N \qquad (5.56)$$

where $p(\widetilde{x}) = p(x_1,\dots,x_N) = p(x(t_1),\dots,x(t_N))$.

An approximation of the amount of information per sample can be found by dividing $H(\widetilde{X})$ by $N = 2wT$.

In general the result is not equal to $H(X)$, it being the amount of information in just one sample. For information sources without memory only, it is the case that

$$\frac{H(\widetilde{X})}{N} = H(X). \qquad (5.57)$$

In that case

$$p(x_1,x_2,\dots,x_N) = \prod_{i=1}^{N} p(x_i)$$

and thus

$$H(\widetilde{X}) = -\int_{-\infty}^{\infty} \dots \int_{-\infty}^{\infty} p(x_1,x_2,\dots,x_N) \log p(x_1,x_2,\dots,x_N) \, dx_1 dx_2 \dots dx_N$$

$$= \sum_{i=1}^{N} \left\{ -\int_{-\infty}^{\infty} p(x_i) \log p(x_i) \, dx_i \right\} = N H(X).$$

Clearly, the quantity $H(\widetilde{X})/N$ corresponds with $H_N(U)$ in Section 3.2. In analogy to equation (5.48) we can define the conditional amount of information of a sample, given the previous sample, as

$$H(X_2/X_1) = -\int_{-\infty}^{\infty} \int_{-\infty}^{\infty} p(x_1,x_2) \log p(x_2/x_1) \, dx_1 dx_2. \tag{5.58}$$

This conditional amount of information now also has the property that

$$H(X_2/X_1) \le H(X_2), \tag{5.59}$$

and thus $\quad H(X_1,X_2) \quad = H(X_1) + H(X_2/X_1)$

$$= H(X_2) + H(X_1/X_2) \le H(X_1) + H(X_2), \tag{5.60}$$

with equality if the samples are statistically independent (memoryless source). Knowledge of x_1 therefore leads to a decrease of the uncertainty about x_2. If we consider $2NT$ samples we define, in accordance with Section 3.2,

$$F_N(X) = H(X_N/X_{N-1}, \dots ,X_1) \quad \text{bits/sample,} \tag{5.61}$$

and

$$H_N(X) = \frac{1}{N} H(X_1,\dots,X_N) = \frac{1}{N} H(\widetilde{X}) \quad \text{bits/sample,} \tag{5.62}$$

which is the amount of information per sample where the interdependence of the samples has been taken into account.

If the probability density is gaussian, the various expressions can be worked out further. For other probability densities this is often very difficult.

The amount of information of a stationary gaussian source is given by

$$H(\widetilde{X}) = N \log\left\{ \sigma\sqrt{2\pi e} \right\} \tag{5.63}$$

if the signal is made up of N uncorrelated samples (see equation (5.44)). For dependent samples with N-dimensional probability density $p(\widetilde{x})$ we find

$$H(\widetilde{X}) = \log\left\{ (2\pi e)^{N/2} |\mathbf{K}_{xx}|^{\frac{1}{2}} \right\}, \tag{5.64}$$

where $|\mathbf{K}_{xx}|$ is the determinant of the autocovariance matrix $\mathbf{K}_{xx}$. If the samples are uncorrelated, thus if $\mathbf{K}_{xx}$ is a diagonal matrix with diagonal elements $\sigma 2$, the previously mentioned case follows directly. In analogy with the one-dimensional case, one can show that an information source with a given autocovariance matrix delivers a maximum amount of information if the source is gaussian.

Example 5.2
Assume $N = 2$ and a two-dimensional gaussian probability density function with $\mu_1 = \mu_2 = 0$, $\sigma_1 = \sigma_2 = 1$ and $\rho = \frac{1}{2}$. With the help of equation (5.19) we obtain

$$p(x_1,x_2) = \frac{1}{\pi\sqrt{3}} \, e^{-\frac{2}{3}(x_1{}^2 - x_1 x_2 - x_2{}^2)},$$

and since

$$p(x_1) = \frac{1}{\sqrt{2\pi}} \, e^{-\frac{1}{2}x_1{}^2}$$

the conditional probability density function becomes

$$p(x_2/x_1) = \frac{p(x_1,x_2)}{p(x_1)} = \frac{1}{\frac{1}{2}\sqrt{6\pi}} \, e^{-\frac{2}{3}(x_2 - \frac{1}{2}x_1)^2}.$$

This is again a gaussian distribution; now with $\mu = \frac{1}{2}x_1$ and $\sigma = \frac{1}{2}\sqrt{3}$, and thus

$$H(X_2/X_1) = \log \sigma\sqrt{2\pi e} = \log \frac{1}{2}\sqrt{6\pi e} \approx 1.84,$$

$H(\tilde{X})$ becomes

$$H(\tilde{X}) = H(X_1,X_2) = H(X_1) + H(X_2/X_1) = \log\sqrt{2\pi e} + \log\frac{1}{2}\sqrt{6\pi e}$$

$$= \log(\pi e\sqrt{3}).$$

The same result is obtained if equation (5.64) is applied directly. Since $\sigma_1 = \sigma_2 = 1$ and $\rho = \frac{1}{2}$ it follows with equation (5.27) that $cov(\mathbf{x}_1,\mathbf{x}_2) = \frac{1}{2}$. Using equations (5.34) and (5.35) the autocovariance matrix is given by

$$K_{xx} = \begin{bmatrix} 1 & \frac{1}{2} \\ \frac{1}{2} & 1 \end{bmatrix}.$$

Thus,

$$H(\widetilde{X}) = \log \left\{ (2\pi e)^{N/2} \, |K_{xx}|^{1/2} \right\} = \log \left\{ 2\pi e \begin{bmatrix} 1 & \frac{1}{2} \\ \frac{1}{2} & 1 \end{bmatrix}^{1/2} \right\}$$

$$= \log (\pi e \sqrt{3}) \approx 3.89. \qquad\qquad \triangle$$

Some remarks with respect to the transformation of stochastic variables will be made at the end of this section. In the case of the transformation of a stochastic variable **x** to a stochastic variable **y**, according to **y** = *f*(**x**), the probability density function will also change and as such the amount of information. In the general case of an *N*-dimensional probability density functions it is the case that

$$q(y_1,\ldots,y_N) = p(x_1,\ldots,x_N) \left| J\left(\frac{x_1\ldots x_N}{y_1\ldots y_N} \right) \right|, \qquad (5.65)$$

whence

$$\int_{x_1}\int_{x_2} \ldots \int_{x_N} p(x_1,\ldots,x_N) dx_1 dx_2 \ldots dx_N = 1,$$

and

$$J\left(\frac{x_1\ldots x_N}{y_1\ldots y_N} \right)$$

is called the Jacobian of $x_1,x_2,\ldots x_N$ with respect to $y_1,y_2,\ldots,y_N$. For $N = 3$ the Jacobian is defined as follows:

$$J\left(\frac{x_1,x_2,x_3}{y_1,y_2,y_3} \right) = \begin{bmatrix} \dfrac{\partial x_1}{\partial y_1} & \dfrac{\partial x_1}{\partial y_2} & \dfrac{\partial x_1}{\partial y_3} \\[2mm] \dfrac{\partial x_2}{\partial y_1} & \dfrac{\partial x_2}{\partial y_2} & \dfrac{\partial x_2}{\partial y_3} \\[2mm] \dfrac{\partial x_3}{\partial y_1} & \dfrac{\partial x_3}{\partial y_2} & \dfrac{\partial x_3}{\partial y_3} \end{bmatrix}. \qquad (5.66)$$

Notice that

$$\left| J\left(\frac{x_1\ldots x_N}{y_1\ldots y_N} \right) \right| = \frac{1}{\left| J\left(\dfrac{y_1\ldots y_N}{x_1\ldots x_N} \right) \right|}. \qquad (5.67)$$

For the amount of information $H(\widetilde{Y})$ we find

$$H(\tilde{Y}) = -\int_{y_1}\int_{y_2} \ldots \int_{y_N} p(x_1,\ldots,x_N) \left| J\left(\frac{x_1\ldots x_N}{y_1\ldots y_N}\right) \right|$$

$$\times \log \left\{ p(x_1\ldots,x_N) \left| J\left(\frac{x_1\ldots x_N}{y_1\ldots y_N}\right) \right| \, \mathrm{d}y_1\ldots\mathrm{d}y_N \right\}.$$

(5.68)

The one-dimensional case yields

$$H(Y) = -\int_y q(y) \log q(y) \, \mathrm{d}y$$

$$= -\int_x p(x)J(x/y) \log \{p(x) J(x/y)\} \, J(y/x) \, \mathrm{d}x$$

$$= H(X) - E_x \{\log J(x/y)\}.$$

(5.69)

Thus the amount of information $H(Y)$ is identical to $H(X)$ except for a constant term.

Special attention must be paid to situations where the probability density function of a stochastic variable should be determined which is e.g. the sum of two other stochastic variables.

Let $\mathbf{x}$ and $\mathbf{y}$ be stochastic variables and $\mathbf{z}$ a stochastic variable such that $\mathbf{z} = \mathbf{x} + \mathbf{y}$. The probability density functions with respect to $\mathbf{x}$ and $\mathbf{y}$ are given by $p(x)$ and $p(y)$ respectively. For the cumulative distribution function it is the case that

$$F(z) = P(x + y \le z) = \int_{-\infty}^{\infty} \int_{-\infty}^{z-x} p(x,y) \, \mathrm{d}x\mathrm{d}y.$$

Therefore

$$p(z) = \frac{\mathrm{d}F(z)}{\mathrm{d}z} = \int_{-\infty}^{\infty} \frac{\mathrm{d}}{\mathrm{d}z} \left\{ \int_{\infty}^{z-x} p(x,y) \, \mathrm{d}y \right\} \mathrm{d}x.$$

Since, in general,

$$\frac{\mathrm{d}}{\mathrm{d}u} \int_{-\infty}^{u} f(r) \, \mathrm{d}r = f(u) - f(-\infty),$$

(5.70)

and $\mathbf{x}$ and $\mathbf{y}$ are independent, it is the case that

$$p(z) = \int\limits_{-\infty}^{\infty} p(x,z-x) \, \mathrm{d}x = \int\limits_{-\infty}^{\infty} p(x)p_y(z-x) \, \mathrm{d}x. \tag{5.71}$$

Example 5.3
Let

$$p(x) = \frac{1}{2}, \ 0 \le x \le 2,$$

$$p(y) = \frac{1}{2}, \ 0 \le y \le 2,$$

and $\mathbf{z} = \mathbf{x} + \mathbf{y}$.
Since $0 \le y \le 2$, $0 \le z - x \le 2$ and thus $z - 2 \le x \le z$, two cases for $p(z)$ are distinguished.

(a) $z \le 2$

Since both x and y are larger than 0, also $z \ge 0$. Thus $0 \le z \le 2$.
For x it follows that $0 \le x \le z$ and $z - 2 \le x \le z$. See Figure 5.13(a).
Combination leads to $0 \le x \le z$. For $p(z)$ we find

$$p(z) = \int\limits_{0}^{z} \frac{1}{2}\frac{1}{2} \, \mathrm{d}x = \frac{1}{4}x \, \bigg|_{0}^{z} = \frac{1}{4}z.$$

(b) $z \ge 2$

Since the maximum values of x and y are 2, $2 \le z \le 4$.
Now $z - 2 \le x \le 2$ (see Figure 5.13(b)) and thus

$$p(z) = \int\limits_{z-2}^{2} \frac{1}{2}\frac{1}{2} \, \mathrm{d}x = \frac{1}{4}x \, \bigg|_{z-2}^{2} = 1 - \frac{1}{4}z.$$

Figure 5.13.

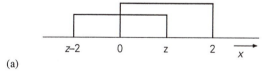

(a)

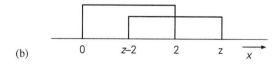

(b)

For $H(Z)$ we obtain

$$H(Z) = -\int_0^2 \tfrac{1}{4} z \log \tfrac{1}{4} z \, dz - \int_2^4 (1 - \tfrac{1}{4} z) \log (1 - \tfrac{1}{4} z) \, dz$$

$$= -4 \int_0^{\frac{1}{2}} u \log u \, du + 4 \int_{\frac{1}{2}}^0 u \log u \, du$$

$$= -8 \int_0^{\frac{1}{2}} u \log u \, du = -\frac{8}{\ln 2} u^2 (\tfrac{1}{2} \ln u - \tfrac{1}{4}) \Big|_0^{\frac{1}{2}}$$

$$= 1 + \tfrac{1}{2} \log e. \qquad\qquad \triangle$$

5.5 Information power

In Definition 5.1 of Section 5.3 the amount of information was given for a continuous memoryless source. As seen in the same section the maximum amount of information that a source can deliver depends on the constraints. If for example the average power of the signal (variance) generated by the source is prespecified it follows that max $H(X) = \log[\sigma \sqrt{2\pi e}]$. This maximum value is achieved if the probability density function $p(x)$ of the continuous information source is gaussian and has a variance of σ^2. As mentioned earlier, the gaussian information source is of great importance for describing continuous stochastic signals. It, together with the binary discrete information source, is among the most used models in information theory. In view of the fact that the maximum amount of information depends on the constraints that are given, the concept of redundancy for a continuous source should be dealt with carefully. One can only speak of the redundancy, if the assumed constraints are taken into account. In the case of limited powers the redundancy can be defined as

$$\text{red} = 1 - \frac{H(X)}{\log[\sigma\sqrt{2\pi e}]}. \qquad (5.72)$$

Many sources exist which are bounded in power, but which are not gaussian. One can now ascertain how much power a gaussian source must have in comparison to an arbitrary source, in the case where both have the same amount of information. This leads to the concept of *information power.*

The *information power* P_H of a stochastic signal $x(t)$ generated by an information source is equal to the power of a gaussian signal that has the same amount of information as that stochastic signal.

Let $H(X)$ be the amount of information related to the stochastic signal $x(t)$. According to the definition of information power that value of σ^2 is searched for which makes the amount of information of a gaussian source with that value of σ^2 equal to $H(X)$. The information power P_H is then the same as that of σ^2. From

$$H(X) = \log \sigma \sqrt{2\pi e} = \log \sqrt{2\pi e P_H} \qquad (5.73)$$

it follows that

$$P_H = \frac{1}{2\pi e} 2^{2H(X)}, \qquad (5.74)$$

which is the mathematical expression of information power in the case of arbitrary signals with an amount of information $H(X)$.

Since a gaussian signal leads to a maximum amount of information for a given power, the information power of an arbitrary signal will always be smaller than for the gaussian case.

Example 5.4
Let a probability density function $p(x)$ be given by

$$p(x) \ = \frac{x^3}{2500} \qquad \text{for } 0 \leq x \leq 10,$$
$$= 0 \qquad \text{otherwise.}$$

The amount of information $H(X)$ equals

$$H(X) \ = - \int_{-\infty}^{\infty} p(x) \log p(x) \, dx$$

$$= -\int_{0}^{10} \frac{x^3}{2500} \log \frac{x^3}{2500} \, dx$$

$$= -\frac{3}{2500} \int_{0}^{10} x^3 \log x \, dx + \frac{\log 2500}{2500} \int_{0}^{10} x^3 \, dx$$

$$= -\frac{3 \log e}{2500} \int_{0}^{10} x^3 \ln x \, dx + \frac{\log 2500}{2500} \int_{0}^{10} x^3 \, dx$$

$$= -\frac{3\log e}{2500}\left[\frac{x^4}{4}\left(\ln x - \frac{1}{4}\right)\right]_0^{10} + \frac{\log 2500}{2500}\left[\frac{1}{4}x^4\right]_0^{10}$$

$$= \log\frac{5}{2}\sqrt[4]{e^3} \approx 2.40.$$

For the mean and variance it follows, respectively, that

$$\mu = \int_0^{10} \frac{x^4}{2500}\,dx = \left[\frac{x^5}{12500}\right]_0^{10} = 8,$$

$$\sigma^2 = \int_0^{10}(x-8)^2\frac{x^3}{2500}\,dx = \frac{1}{2500}\int_0^{10}x^5\,dx - \frac{64}{2500}\int_0^{10}x^3\,dx$$

$$= \left[\frac{x^6}{15000} - \frac{16x^4}{2500}\right]_0^{10} = \frac{8}{3}.$$

Given $\sigma^2 = \frac{8}{3}$ the maximum amount of information is equal to $\log \sigma \sqrt{2\pi e}$

$$= \log\sqrt{2\pi e \frac{8}{3}} \approx 2.75.$$

As a result the redundancy of the continuous source becomes

$$\text{red} = 1 - \frac{H(X)}{\log \sigma\sqrt{2\pi e}} = 1 - \frac{2.40}{2.75} = 0.13.$$

For the information power we find

$$P_H = \frac{1}{2\pi e}2^{2H(X)} = \frac{1}{2\pi e}\left\{\frac{5}{2}\sqrt[4]{e^3}\right\}^2 = \frac{25\sqrt{e}}{8\pi} \approx 1.64.$$

Clearly, the information power is indeed less than the actual power which is equal to $P = \sigma^2 = \frac{8}{3} \approx 2.66$. △

The concept of information power is of importance when describing a continuous communication channel where noise with a non-gaussian probability density is present.

In the general case where we are dealing with the sum $z(t)$ of two signals $x(t)$ and $y(t)$, $z(t) = x(t) + y(t)$, it can be shown that the following inequality holds:

$$P_{H_x} + P_{H_y} \le P_{H_z} \le P_x + P_y, \tag{5.75}$$

where P_{H_x}, P_{H_y} and P_{H_z} are the information powers of the two signals $\mathbf{x}(t)$, $\mathbf{y}(t)$ and $\mathbf{z}(t)$, respectively. P_x and P_y are the powers of $\mathbf{x}(t)$ and $\mathbf{y}(t)$, repectively. The equals sign in the above expression holds if both signals are gaussian because in this case information power and actual power are the same.

What the inequality actually says is that if two signals are superimposed the amount of information is increased or at the very least not decreased by the superposition. This follows by expressing the (information) power in terms of information measures. That the amount of information will increase or at the very least not decrease in case of superposition can be elucidated as follows. If two independent signals $\mathbf{x}(t)$ and $\mathbf{y}(t)$ are superimposed, it may be expected that their superposition $\mathbf{z}(t)$ would tend to become a random-noise signal. Since random noise is characterized by a gaussian distribution this means that the probability density function of $\mathbf{z}(t)$ tends toward a gaussian distribution. The fact that gaussian distributions lead to a maximum amount of information has as consequence that in the case of superposition an increase of the amount of information may be expected.

Example 5.5
Suppose we have two continuous information sources with the following probability density functions.

$$p(x) = \frac{1}{2} \quad \text{for } 0 \le x \le 2,$$
$$= 0 \quad \text{otherwise.}$$

$$p(y) = \frac{1}{2} \quad \text{for } 0 \le x \le 2,$$
$$= 0 \quad \text{otherwise.}$$

This corresponds to $H(X) = H(Y) = 1$ and thus

$$P_{H_x} = P_{H_y} = \frac{1}{2\pi \, e} \, 2^{2H(X)} = \frac{2}{\pi \, e}.$$

The actual power is $P_x = P_y = \frac{1}{3}$, as can be computed from σ_x^2.
If $\mathbf{x}$ and $\mathbf{y}$ are independent the probability density function with respect to $\mathbf{z} = \mathbf{x} + \mathbf{y}$ is given by (compare Example 5.3)

$$p(z) = \frac{1}{4} z \quad \text{for } 0 \le z \le 2,$$
$$= 1 - \frac{1}{4} z \quad \text{for } 2 \le z \le 4,$$
$$= 0 \quad \text{otherwise.}$$

If $H(Z)$ is calculated one finds

and
$$H(Z) = \log 2\sqrt{e},$$

$$P_z = \sigma_z^2 = \frac{2}{3}.$$

For the information it is the case that

$$P_{H_z} = \frac{1}{2\pi e} \, 2^{2 \log 2\sqrt{e}} = \frac{2}{\pi} \approx 0.64.$$

If we compare this with the inequality of formula (5.75) we find

$$P_{H_x} + P_{H_y} = \frac{4}{\pi e} = 0.47 \le P_{H_z} = \frac{2}{\pi} = 0.64 \le P_x + P_y = \frac{2}{3} = 0.67.$$

$\triangle$

5.6 Exercises

5.1. *a.* The stochastic variable **x** represents the amplitude of a signal $x(t)$ that is bounded between -3 and $+3$ V. **x** has a uniform probability distribution between these limits. Determine the amount of information $H(X)$.

b. Determine the amount of information $H(X)$ if **x** is uniformly distributed between -5 and $+5$ V.

c. Give an explanation for the difference between the answers found in a and b.

5.2. A sample **x** of a stochastic signal is uniformly distributed between $+1$ and $+7$ V.

a. Determine the amount of information $H(X)$. What conclusions can you make if you compare this result with that of Exercise 5.1.a?

b. Determine $E(\mathbf{x})$ and var($\mathbf{x}$).

5.3. Given is the continuous stochastic variable **x** with probability density $p(x)$. For the stochastic variable **y** we have

$$\mathbf{y} = \mathbf{x} + \alpha.$$

Prove that $H(X) = H(Y)$.

5.4. The stochastic variable **x** has a probability density as given in Figure 5.14.

a. Determine the value of k.

b. Determine $H(X)$.

c. Compare the value found for $H(X)$ with the answer of Exercise 5.1.a and give an explanation for any difference.

5.5. A sample **x** of the stochastic signal **x**(*t*) is uniformly distributed between $-A$ and $+A$ with $A > 0$.

a. Sketch $H(X)$ as a function of A.

b. Determine and sketch var(**x**) as a function of A.

c. What do the two graphs have in common?

5.6. The stochastic variable **x** has a negative exponential distribution:

$$p(x) = \frac{1}{\lambda} \exp(-\frac{x}{\lambda}) \qquad \text{for } x \geq 0,$$

$$= 0 \qquad \text{for } x < 0.$$

a. Calculate the amount of information $H(X)$.

b. Consider a stochastic variable that cannot assume negative values and of which the expectation is λ. Show that this variable has a maximum amount of information when its probability density is given by the above distribution.

5.7. A sample **x** of a signal $x(t)$ has the following probability density:

$$p(x) = \frac{1}{2\lambda} \exp\left(-\frac{|x|}{\lambda}\right).$$

For a sample **y** of the output signal we have

$$\mathbf{y} = |\mathbf{x}|.$$

a. Determine $H(X)$.

b. Determine $q(y)$ and from this, determine $H(Y)$.

c. Reason out that $H(Y/X) = 0$.

Figure 5.14. Probability density $p(x)$ of Exercise 5.4.

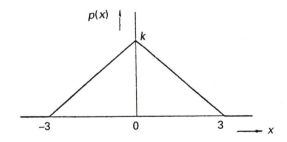

d. Determine $H(X/Y)$ from the previous answers. Can you give an explanation for the values found?

5.8. The statistically independent variables **x** and **y**, which represent the signals $x(t)$ and $y(t)$, have the probability densities shown in Figure 5.15.

Figure 5.15. Probability density $p(x)$ and $p(y)$ of Exercise 5.8.

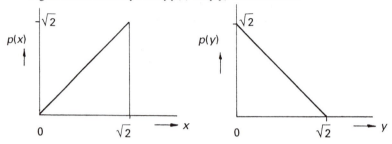

The signals $x(t)$ and $y(t)$ are added to a discriminator (see Figure 5.16), where for a sample **z** of the output signal we have

$$\mathbf{z} = \mathbf{x} \qquad \text{for} \qquad \mathbf{x} \geq \frac{1}{2}\sqrt{2},$$

$$\mathbf{z} = \mathbf{y} \qquad \text{for} \qquad \mathbf{x} < \frac{1}{2}\sqrt{2}.$$

Figure 5.16. The discriminator of Exercise 5.8.

a. Determine $H(Z/x < \frac{1}{2}\sqrt{2})$.

b. Determine $H(Z/x \geq \frac{1}{2}\sqrt{2})$.

c. Determine $H(Z)$.

For b you should keep in mind that $p(x) \neq p(x/x \geq \frac{1}{2}\sqrt{2})$.

5.9. The stochastic variables **x** and **y** have the joint probability density sketched in Figure 5.17.

a. Determine $H(X)$.

b. Determine $H(Y/X)$.

c. Determine the quantity $H(X/Y)$ from a and b.

Check your answer by calculating $H(X/Y)$ directly.

d. On the basis of your answer, conclude if **x** and **y** are statistically independent.

5.10. For the stochastic variables **x** and **y** it is given that:

- $p(x,y)$ is the two-dimensional probability density;

- $E(\mathbf{x}) = E(\mathbf{y}) = 0$;

- $\mathrm{var}(\mathbf{x}) = \sigma_1^2$ and $\mathrm{var}(\mathbf{y}) = \sigma_2^2$;

- the correlation coefficient is ρ.

a. Determine $H(Y)$.
b. Determine $H(X/Y)$.
c. Determine $H(X,Y)$.
d. How are these quantities influenced if $\rho = 0$ and how do you explain this?

5.11. A continuous information source generates a stochastic signal $\mathbf{x}(t)$ with the following probability density:

$$p(x) = e^{-a|x|}.$$

a. Calculate the amount of information in a sample of this signal.
b. What does one understand by information power?
c. Calculate the information power for the above signal.

5.12. A continuous information source generates a stochastic signal $\mathbf{x}(t)$ with the following probability density:

$$p(x) = ax^2, \qquad 0 \le x \le \lambda.$$

Figure 5.17. Joint probability density of Exercise 5.9.

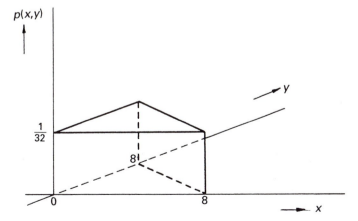

a. Calculate the amount of information $H(X)$ for one sample and sketch $H(X)$ as a function of λ ($\lambda > 0$).

b. For what value of λ is $H(X) = 0$? Does this value of λ have any special meaning?

c. Calculate the information power of this source.

5.13. A continuous information source generates a stochastic signal $x(t)$ with the following probability density:

$$p(x) = \frac{1 - \dfrac{|x|}{a}}{a} \qquad \text{for} \quad |x| \le a,$$

$$= 0 \qquad\qquad \text{for} \quad |x| > a.$$

a. Calculate the amount of information in a sample of this signal.

b. Calculate the information power of the source.

c. Find an expression for the power of the source as a function of a ($a > 0$).

d. Compare $H(X)$ with the amount of information $H(Y)$ of a gaussian source that has the same power as the source under consideration.

5.7 Solutions

5.1. *a.* Generally, for a stochastic variable x which is bounded between $-A$ and $+A$, while it has a uniform distribution between these bounds, we have that

$$p(x) = \frac{1}{2A} \quad \text{for} \quad -A \le x \le A.$$

On the basis of Theorem 5.2 the corresponding amount of information is equal to

$$H(X) \;=\; \log 2A.$$

In this case $A = 3$, hence

$$H(X) = \log 6 = 2.58 \;\; \text{bits.}$$

b. Substitution of $A = 5$ into the general equation given in a leads to

$$H(X) = \log 10 = 3.32 \;\; \text{bits.}$$

c. It can be concluded that the amount of information will increase if the range of the variable is larger. This is in agreement with the fact that the uncertainty regarding the variable increases for a larger range.

5.2. *a.* Since we are dealing with a uniform distribution we have that $p(x) = \frac{1}{6}$ for $1 \le x \le 7$. The amount of information is equal to

$$H(X) = -\int_1^7 p(x) \log p(x)\, dx = -\int_1^7 \frac{1}{6} \log\frac{1}{6}\, dx = \log 6 = 2.58 \text{ bits.}$$

A comparison with the result of Exercise 5.1.a shows that the amount of information for a uniform distribution only depends on the range of the variable, and not on the position of the range.

b. We have

$$E(x) = \int_1^7 x\, p(x)\, dx = \int_1^7 \frac{x}{6}\, dx = \frac{1}{2} x^2 \Big|_1^7 = 4.$$

For the variance we have $\text{var}(\mathbf{x}) = E[(x - E(\mathbf{x}))^2]$.
Since $E(\mathbf{x}) = 4$, it follows that

$$\text{var}(\mathbf{x}) = E[(x-4)^2] = \int_1^7 (x-4)^2 p(x)\, dx$$

$$= \int_1^7 \frac{(x-4)^2}{6}\, dx = \int_{-3}^3 \frac{y^2}{6}\, dy = \frac{1}{18} y^3) \Big|_{-3}^3 = 3.$$

5.3. Since α is a constant, the addition of α to $\mathbf{x}$ will not alter the uncertainty in $\mathbf{y} = \mathbf{x} + \alpha$. The amount of information will therefore also remain unaltered and we will have that $H(X) = H(Y)$. This can also be seen by bearing in mind that the uncertainty with regard to $\mathbf{y}$ is in fact equal to the uncertainty with regard to $\mathbf{x}$ and α: $H(X,\alpha)$.
Since α is a constant the value of which is determined a priori, we have that $H(\alpha) = 0$. It now follows that

$$H(Y) = H(X,\alpha) = H(X) + H(\alpha) = H(X).$$

5.4. *a.* We must have that

$$\int_{-\infty}^{\infty} p(x)\, dx = 1.$$

It can readily be seen that k must then be equal to $\frac{1}{3}$.
b. In order to be able to calculate $H(X)$, we must first find $p(x)$. It can be verified that

$$p(x) = \frac{x+3}{9} \quad \text{for } x \le 0,$$

and

$$p(x) = \frac{3-x}{9} \quad \text{for } x > 0.$$

For $H(X)$ this leads to

$$H(X) = -\int_{-3}^{0} \frac{x+3}{9} \log\left(\frac{x+3}{9}\right) dx - \int_{0}^{3} \frac{3-x}{9} \log\left(\frac{3-x}{9}\right) dx$$

$$= -18 \int_{0}^{1/3} y \cdot \log y \, dy.$$

Since in general

$$\int y \ln y \, dy = y^2 \left[\frac{1}{2} \ln y - \frac{1}{4}\right],$$

it follows for $H(X)$ that

$$H(X) = -\frac{18}{\ln 2} \int_{0}^{1/3} y \ln y \, dy = -\frac{18}{\ln 2} y^2 \left(\frac{1}{2} \ln y - \frac{1}{4}\right)\Big|_{0}^{1/3} =$$

$$= -\log \frac{1}{3} + \frac{1}{2} \log e = 2.30 \text{ bits.}$$

c. The value found here is lower than that found in Exercise 5.1.a. This has to do with the fact that the probability density in this exercise has a peak, causing the uncertainty to decrease, and along with it the amount of information.

5.5. *a.* From Exercise 5.1 there follows $H(X) = \log 2A$. The graph is given in Figure 5.18.

Figure 5.18. $H(X)$ as function of A (Exercise 5.5).

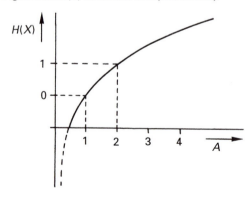

b. var($\mathbf{x}$) can be calculated as follows:

$$E(\mathbf{x}) = \int_{-A}^{A} x\, p(x)\, dx = \int_{-A}^{A} \frac{1}{2A}\, x\, dx = \frac{1}{4A}\, x^2 \Big|_{-A}^{A} = 0.$$

Using this, var($\mathbf{x}$) is then equal to

$$\mathrm{var}(\mathbf{x}) = E[(x - E(\mathbf{x}))^2] = E[x^2] = \int_{-A}^{A} x^2 p(x)\, dx = \frac{1}{6A}\, x^3 \Big|_{-A}^{A} = \frac{1}{3} A^2.$$

See Figure 5.19.

Figure 5.19. Variance as function of *A* (Exercise 5.5).

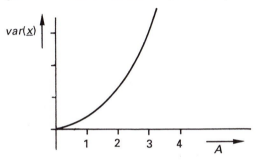

c. The larger *A* becomes the larger the variance and the larger the uncertainty regarding **x**. The latter corresponds with the increase of *H*(*X*) as is shown in the graph of *H*(*X*) as a function of *A*.

5.6. *a.*
$$H(X) = -\int_0^\infty p(x) \log p(x)\, dx = -\log e \int_0^\infty p(x) \ln p(x)\, dx$$

$$= -\log e \int_0^\infty \frac{1}{\lambda} \exp\left(-\frac{x}{\lambda}\right) \ln\left[\frac{1}{\lambda} \exp\left(-\frac{x}{\lambda}\right)\right] dx$$

$$= \log e \frac{\ln \lambda}{\lambda} \int_0^\infty \exp\left(-\frac{x}{\lambda}\right) dx + \log e \frac{1}{\lambda^2} \int_0^\infty x \exp\left(-\frac{x}{\lambda}\right) dx$$

$$= \log e \ln \lambda \int_0^\infty e^{-y} dy + \log e \int_0^\infty y e^{-y} dy =$$

$$= -\log \lambda e^{-y} \Big|_0^\infty + \log e \cdot e^{-y}(-y-1) \Big|_0^\infty = \log \lambda + \log e.$$

b. A probability density *p*(*x*) should be determined in such a way that

$$H(X) = -\int_0^\infty p(x) \log p(x)\, dx$$

is maximum, where the following requirements must be met:

$$\int_0^\infty p(x)\,dx = 1 \qquad \text{and} \qquad \int_0^\infty x\,p(x)\,dx = \lambda.$$

We now form the function

$$G(x) = -p(x)\log p(x) + \alpha_1\,p(x) + \alpha_2\,x\,p(x).$$

If we set the derivative of $G(x)$ with repsect to $p(x)$ equal to zero, then we find

$$-\log p(x) - \log e + \alpha_1 + \alpha_2\,x = 0.$$

After dividing by $-\log e$ this expression becomes

$$\ln p(x) + 1 - \lambda_1 - \lambda_2\,x = 0,$$

where $\lambda_1 = \alpha_1 / \log e$ and $\lambda_2 = \alpha_2 / \log e$. This gives the following as solution:

$$p(x) = e^{\lambda_1 - 1}\,e^{\lambda_2 x}.$$

Substitution of this $p(x)$ into both boundary requirements yields

$$\int_0^\infty p(x)\,dx = \int_0^\infty e^{\lambda_1 - 1}\,e^{\lambda_2 x}\,dx = e^{\lambda_1 - 1}\,e^{\lambda_2 x}/\lambda_2 \Big|_0^\infty = -e^{\lambda_1 - 1}/\lambda_2 = 1,$$

where it is assumed that $\lambda_2 < 0$. Therefore $e^{\lambda_1 - 1} = -\lambda_2$.
We must also have that

$$\int_0^\infty x\,p(x)\,dx = \int_0^\infty x\,e^{\lambda_1 - 1}\,e^{\lambda_2 x}\,dx = -\int_0^\infty x\,\lambda_2\,e^{\lambda_2 x}\,dx = \lambda,$$

from which it follows that

$$\lambda = -\int_0^\infty x\,\lambda_2\,e^{\lambda_2 x}\,dx = -\lambda_2\frac{e^{\lambda_2 x}}{(\lambda_2)^2}(\lambda_2 x - 1)\Big|_0^\infty = -\frac{1}{\lambda_2}.$$

It therefore follows that $\lambda_2 = -1/\lambda$ and $p(x)$ thus becomes

$$p(x) = \frac{1}{\lambda}\exp(\frac{-x}{\lambda}),$$

which is the negative exponential distribution.

5.7. *a.* $H(X) = -\int_{-\infty}^\infty \frac{1}{2\lambda}\exp(-\frac{|x|}{\lambda})\,\log[\frac{1}{2\lambda}\exp(-\frac{|x|}{\lambda})]\,dx$

$$= -\int_{-\infty}^{\infty} \frac{1}{2\lambda} \exp\left(-\frac{|x|}{\lambda}\right) \log\frac{1}{2} \, dx - \int_{-\infty}^{\infty} \frac{1}{2\lambda} \exp\left(-\frac{|x|}{\lambda}\right) \log\left[\frac{1}{\lambda} \exp\left(-\frac{|x|}{\lambda}\right)\right] dx$$

$$= 1 - \int_{0}^{\infty} \frac{1}{\lambda} \exp\left(\frac{-x}{\lambda}\right) \log\left[\frac{1}{\lambda} \exp\left(\frac{-x}{\lambda}\right)\right] dx.$$

If we consider Exercise 5.6.a then it can be concluded that the term with the integral is exactly the amount of information of the negative exponential distribution. In the case now being considered we therefore have for $H(X)$ that

$$H(X) = 1 + \log \lambda + \log e.$$

b. Since **y** = |**x**| the probability of *y* will be equal to the probability of $-x$ and $+x$. Since $p(x) = p(-x)$ the probability of *y* will be equal to twice that of $p(x)$. In other words,

$$q(y) = \frac{1}{\lambda} \exp\left(\frac{-y}{\lambda}\right), \qquad y \ge 0.$$

If we compare this result with Exercise 5.6.a it then directly follows that

$$H(Y) = \log \lambda + \log e.$$

c. If **x** is known, **y** is also known. That is, if **x** is known, there is no uncertainty about **y** and we will therefore have $H(Y/X) = 0$.

d. In general, we have

$$H(X,Y) = H(X) + H(Y/X) = H(Y) + H(X/Y).$$

If one substitutes the previously found values for $H(X)$, $H(Y)$ and $H(Y/X)$ it then follows that

$$H(X/Y) = 1.$$

That is, if **y** is known, there is still some uncertainty over **x**. This is indeed the case since the sign is unknown. The value of **x** can be positive or negative.

5.8. *a.* If $x < \frac{1}{2}\sqrt{2}$ then **z** = **y**. Therefore

$$H(Z/x < \tfrac{1}{2}\sqrt{2}) = H(Y/x < \tfrac{1}{2}\sqrt{2}) = H(Y).$$

For the probability density $p(y)$ we have $p(y) = \sqrt{2} - y$ for $0 \le y \le \sqrt{2}$. The amount of information can now be calculated as follows:

$$H(Y) = -\int_0^{\sqrt{2}} (\sqrt{2} - y) \log(\sqrt{2} - y) \, dy$$

$$= \int_{\sqrt{2}}^0 t \log t \, dt = \frac{1}{\ln 2} t^2 \left(\frac{1}{2} \ln t - \frac{1}{4}\right) \Big|_{\sqrt{2}}^0$$

$$= -\log \sqrt{2} + \frac{1}{2 \ln 2} = \frac{1}{2} \log \frac{e}{2} \text{ bit.}$$

b. Since $z = x$ for $x \geq \frac{1}{2} \sqrt{2}$ we have that

$$H(Z/x \geq \frac{1}{2}\sqrt{2}) = H(X/x \geq \frac{1}{2}\sqrt{2}).$$

From the Bayes formula it follows that

$$p(x/x \geq \frac{1}{2}\sqrt{2}) = \frac{p(x) \cdot p(x \geq \frac{1}{2}\sqrt{2}/x)}{p(x \geq \frac{1}{2}\sqrt{2})}$$

$$= \frac{p(x)}{p(x \geq \frac{1}{2}\sqrt{2})} = \frac{4}{3} p(x) \quad \text{for} \quad \frac{1}{2}\sqrt{2} < x < \sqrt{2}.$$

From this, together with $p(x) = x$, it follows that

$$H(X/x \geq \frac{1}{2}\sqrt{2}) = -\int_{\frac{1}{2}\sqrt{2}}^{\sqrt{2}} \frac{4}{3} p(x) \log\left\{\frac{4}{3} p(x)\right\} dx = -\int_{\frac{1}{2}\sqrt{2}}^{\sqrt{2}} \frac{4}{3} x \log\left\{\frac{4}{3} x\right\} dx$$

$$= -\int_{\frac{2}{3}\sqrt{2}}^{\frac{4}{3}\sqrt{2}} \frac{3}{4} t \log t \, dt = -\frac{3}{4 \ln 2} t^2 \left(\frac{1}{2} \ln t - \frac{1}{4}\right) \Big|_{\frac{2}{3}\sqrt{2}}^{\frac{4}{3}\sqrt{2}}$$

$$= -\frac{8}{3}\left[\frac{1}{2}\log(4\sqrt{2}) - \frac{1}{2}\log 3 - \frac{1}{4}\log e\right] + \frac{2}{3}\left[\frac{1}{2}\log(2\sqrt{2}) - \frac{1}{2}\log 3 - \frac{1}{4}\log e\right]$$

$$= -\frac{16}{7} + \log(3\sqrt{e}) \text{ bits.}$$

c. $H(Z) = p(x < \frac{1}{2}\sqrt{2}) \cdot H(Z/x < \frac{1}{2}\sqrt{2}) + p(x \geq \frac{1}{2}\sqrt{2}) \cdot H(Z/x > \frac{1}{2}\sqrt{2})$.
Since $p(x < \frac{1}{2}\sqrt{2}) = \frac{1}{4}$ and $p(x \geq \frac{1}{2}\sqrt{2}) = \frac{3}{4}$ it follows together with the previously found values in a and b that

$$H(Z) = \frac{1}{4} \cdot \frac{1}{2}\log\frac{e}{2} + \frac{3}{4}\left[-\frac{17}{6} + \log(3\sqrt{e})\right] = -\frac{1}{2}\log e - \frac{9}{4} + \frac{3}{4}\log 3.$$

5.9. *a.* For $p(x,y)$ we have

$$p(x,y) = \frac{1}{32} \quad \text{for} \quad 0 \le x \le 8 \quad \text{and} \quad 0 \le y \le 8\text{--}x.$$

The probability density $p(x)$ can be calculated as follows:

$$p(x) = \int_0^{8-x} p(x,y)\, dy = \frac{8-x}{32}.$$

The amount of information $H(X)$ is then equal to

$$H(X) = -\int_0^8 \frac{8-x}{32} \log\left(\frac{8-x}{32}\right) dx = \int_{1/4}^0 32\, t \log t \, dt$$

$$= 32\frac{t^2}{\ln 2}\left(\frac{1}{2}\ln t - \frac{1}{4}\right)\Big|_{1/4}^0 = 2 + \frac{1}{2}\log e.$$

b. We have $p(y/x) = p(x,y)/p(x) = 1/(8 - x)$, where $0 < y < 8 - x$. It now follows for the conditional amount of information that

$$H(Y/X) = -\int_0^8 \int_0^{8-x} p(x,y) \log p(y/x)\, dy dx$$

$$= \int_0^8 \int_0^{8-x} \frac{1}{32} \log(8 - x)\, dy dx = \frac{1}{32}\int_0^8 (8 - x)\log(8 - x)\, dx$$

$$= -\frac{1}{32}\int_8^0 t \log t\, dt = -\frac{1}{32}\frac{t^2}{\ln 2}\left(\frac{1}{2}\ln t - \frac{1}{4}\right)\Big|_8^0 = 3 - \frac{1}{2}\log e.$$

c. With the help of the answers to a and b it follows that

$$H(X,Y) = H(X) + H(Y/X) = 2 + \frac{1}{2}\log e + 3 - \frac{1}{2}\log e = 5 \quad \text{bits.}$$

Direct calculation yields

$$H(X,Y) = -\int_0^8 \int_0^{8-x} p(x,y)\cdot\log p(x,y)\, dx dy = -\int_0^8 \int_0^{8-x} \frac{1}{32}\log\frac{1}{32}\, dx dy = 5 \quad \text{bits.}$$

5.10. *a.* Since $p(x,y)$ is the two-dimensional gaussian probability density, where $E(\mathbf{y}) = 0$ and $\text{var}(\mathbf{y}) = \sigma_2^2$, it follows for $q(y)$ that

$$q(y) = \frac{1}{\sigma_2\sqrt{2\pi}} \exp\left(-\frac{y^2}{2\sigma_2^2}\right).$$

On the basis of Theorem 5.3 it follows that

$$H(Y) = \log(\sigma_2\sqrt{2\pi\, e}).$$

b. The probability density $p(x/y)$ can be derived from

$$p(x/y) = p(x,y)/q(y)$$

$$= \frac{1}{\sigma_1\sqrt{2\pi}\sqrt{1-\rho^2}}\exp\left[-\frac{1}{2(1-\rho^2)}\left\{\frac{x^2}{\sigma_1^2}-\frac{2\rho xy}{\sigma_1\sigma_2}+\frac{y^2}{\sigma_2^2}\right\}+\frac{y^2}{2\sigma_2^2}\right]$$

$$= \frac{1}{\sqrt{2\pi}\sigma_1\sqrt{1-\rho^2}}\exp\left[-\frac{(x-\frac{\sigma_1}{\sigma_2}\rho y)^2}{2\sigma_1^2(1-\rho^2)}\right].$$

This is also a gaussian probability density with conditional mean $(\sigma_1/\sigma_2)\rho y$ and variance $\sigma_1^2(1-\rho^2)$.

For $H(X/Y)$ it now follows that

$$-\int_{-\infty}^{\infty} q(y)\left\{\int_{-\infty}^{\infty} p(x/y)\log p(x/y)\,dx\right\}dy,$$

where $p(x/y)$ is equal to the guassian probability density given above. Since the amount of information in the case of a gaussian probability density only depends on the variance and not on the mean, as can readily be shown, it follows with the help of Theorem 5.3 that

$$-\int_{-\infty}^{\infty} p(x/y)\log p(x/y)\,dy = \log\left[\sigma_1\sqrt{1-\rho^2}\sqrt{2\pi\, e}\right],$$

and therefore also

$$H(X/Y) = \log\left[\sigma_1\sqrt{1-\rho^2}\sqrt{2\pi\, e}\right].$$

c. Since $H(X,Y) = H(Y) + H(X/Y)$ it follows from the results of a and b that

$$H(X,Y) = \log\left[\sigma_2\sqrt{2\pi\, e}\right] + \log\left[\sigma_1\sqrt{1-\rho^2}\sqrt{2\pi\, e}\right] =$$

$$= \log\left[\sigma_1\sigma_2\sqrt{1-\rho^2}\,2\pi\, e\right].$$

d. If $\rho = 0$ then this means that $(x - E(\mathbf{x}))$ and $(y) - E(\mathbf{y}))$ do not show any interrelation. Since it is given that $E(\mathbf{x}) = E(\mathbf{y}) = 0$ it therefore follows that $\mathbf{x}$ and $\mathbf{y}$ show no interrelation. In other words, $\mathbf{x}$ and $\mathbf{y}$ are independent of each other. Substitution of $\rho = 0$ in $H(X/Y)$ yields

$$H(X/Y) = \log\left[\sigma_1 \sqrt{2\pi\, e}\right],$$

which is identical to $H(X)$. This agrees with the fact that $\mathbf{x}$ and $\mathbf{y}$ are independent for $\rho = 0$. Substitution of $\rho = 0$ in $H(X,Y)$, as is calculated in c, gives

$$H(X,Y) = \log\left[\sigma_1 \sigma_2 2\pi\, e\right] = H(X) + H(Y),$$

which again agrees with the independence of $\mathbf{x}$ and $\mathbf{y}$ implied by $\rho = 0$.

5.11. *a*. The probability density function is $p(x) = e^{-a|x|}$ (see Figure 5.20). The constant a remains to be determined by integrating and letting the integral be equal to 1. This yields in succession

$$\int_{-\infty}^{0} e^{ax}\, dx + \int_{0}^{\infty} e^{-ax}\, dx = 1,$$

$$\rightarrow \quad 2\int_{0}^{\infty} e^{-ax}\, dx = 1,$$

$$\rightarrow \quad -\frac{2}{a} e^{-ax}\Big|_{0}^{\infty} = 1,$$

so that $\quad a = 2 \quad$ and $\quad p(x) = e^{-2|x|}$.

For $H(X)$ we have by definition that

Figure 5.20. The probability density of Exercise 5.11.

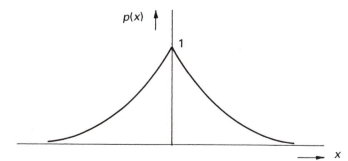

$$H(X) = -\int\limits_{-\infty}^{\infty} p(x)\, \log p(x)\, dx.$$

After substituting $p(x)$ it follows that

$$H(X) = -\int\limits_{-\infty}^{0} e^{2x} \log e^{2x}\, dx - \int\limits_{0}^{\infty} e^{-2x} \log e^{-2x}\, dx.$$

Making use of the symmetry of $p(x)$ it follows that

$$
\begin{aligned}
H(X) \;&= 4 \log e \int\limits_{0}^{\infty} x e^{-2x}\, dx \\
&= -2 \log e \int\limits_{0}^{\infty} x\, de^{-2x} = -2 \; x\, e^{-2x}\log e \,\Big|_{0}^{\infty} + 2 \log e \int\limits_{0}^{\infty} e^{-2x}\, dx \\
&= -\, e^{-2x}\log e \,\Big|_{0}^{\infty} = \log e \quad \text{bits/sample.}
\end{aligned}
$$

b. By the information power P_{H} of a stochastic signal $\mathbf{x}(t)$ generated by an information source, we understand the power of a gaussian signal that has an equally large amount of information as the stochastic signal $\mathbf{x}(t)$.

c. The information power is given by the expression

$$P_{\mathrm{H}} = \frac{1}{2\pi\, e}\, 2^{2H(X)}.$$

Substitution of $H(X)$ gives

$$P_{\mathrm{H}} = \frac{1}{2\pi\, e}\, 2^{2\log e} = \frac{e^2}{2\pi\, e} = \frac{e}{2\pi}.$$

5.12. *a.* See Figure 5.21.

Figure 5.21. The probability density of Exercise 5.12.

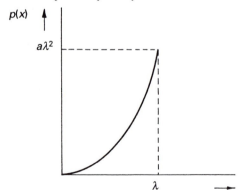

First of all, the parameter a is calculated by integrating and letting the integral assume the value 1. This yields

$$\int_0^\lambda ax^2\,dx = \tfrac{1}{3}ax^3\Big|_0^\lambda = \tfrac{1}{3}a\lambda^3 = 1,$$

so that

$$a = \frac{3}{\lambda^3} \quad \text{and} \quad p(x) = \frac{3}{\lambda^3}x^2.$$

The amount of information in a sample now becomes

$$H(X) = -\int_0^\lambda \frac{3}{\lambda^3}x^2 \log\Big(\frac{3}{\lambda^3}x^2\Big)\,dx = -\frac{6}{\lambda^3}\int_0^\lambda x^2\cdot\log x\,dx - \frac{3}{\lambda^3}\int_0^\lambda x^2\log\frac{3}{\lambda^3}\,dx$$

$$= -\frac{2}{\lambda^3}\int_0^\lambda \log x\,dx^3 - \frac{3}{\lambda^3}\log\frac{3}{\lambda^3}\cdot\frac{1}{3}\lambda^3$$

$$= -\frac{2}{\lambda^3}x^3\log x\,\Big|_0^\lambda + \frac{2}{\lambda^3}\int_0^\lambda x^3\,d\log x - \log\frac{3}{\lambda^3}$$

$$= -2\log\lambda + \frac{2}{\lambda^3}\log e\cdot\frac{1}{3}x^3\,\Big|_0^\lambda - \log\frac{3}{\lambda^3} = -2\log\lambda + \frac{2}{3}\log e - \log\frac{3}{\lambda^3}$$

$$= \log\lambda - \log 3 + \frac{2}{3}\log e = \log\frac{\lambda e^{\frac{2}{3}}}{3} \quad \text{bits/sample.}$$

See Figure 5.22.

Figure 5.22. $H(X)$ as function of λ (Exercise 5.12).

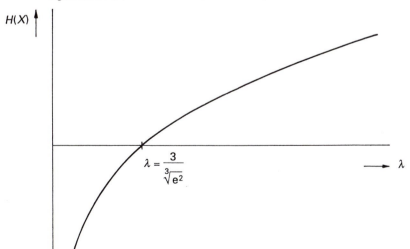

b. From the expression found in a we see that if $H(X) = 0$, we have for λ

$$\lambda = 3e^{-\frac{2}{3}}.$$

$H(X)$ becomes negative for smaller values of λ.
Thus $\lambda = 3e^{-2/3}$ can be regarded as the lower bound where it is still meaningful to work with an accuracy of $\Delta x = 1$.
c. The information power is given by the expression

$$P_H = \frac{1}{2\pi e} \, 2^{2H(X)}.$$

Substitution of $H(X)$ gives

$$P_H = \frac{1}{2\pi e} \, 2^{2\log(\lambda e^{2/3}/3)} = \frac{1}{2\pi e} \left(\frac{\lambda e^{\frac{2}{3}}}{3} \right)^2.$$

5.13. *a*. See Figure 5.23.

Figure 5.23. The probability density of Exercise 5.13.

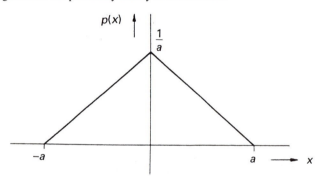

For $H(X)$ we have by definition that

$$H(X) = -\int_{-\infty}^{\infty} p(x) \log p(x) \, dx.$$

After substituting $p(x)$ it follows that

$$H(X) = -\frac{1}{a} \int_{-a}^{0} (1 + \frac{x}{a}) \log\left\{ \frac{1}{a}(1 + \frac{x}{a}) \right\} \, dx - \frac{1}{a} \int_{0}^{a} (1 - \frac{x}{a}) \log\left\{ \frac{1}{a}(1 - \frac{x}{a}) \right\} \, dx.$$

Making use of the symmetry of $p(x)$ it follows that

$$H(X) = -\frac{2}{a}\int_0^a (1-\frac{x}{a}) \log\{\frac{1}{a}(1-\frac{x}{a})\} \, dx$$

$$= -\frac{2}{a}\int_0^a (1-\frac{x}{a}) \log \frac{1}{a} \, dx - \frac{2}{a}\int_0^a (1-\frac{x}{a}) \log(1-\frac{x}{a}) \, dx.$$

For the first integral after the equals sign it now follows that

$$= -2\log a \int_0^a (1-\frac{x}{a}) \, d(1-\frac{x}{a}) = -2\log a \cdot \frac{1}{2} (1-\frac{x}{a})^2 \Big|_0^a$$

$$= -2\log a - \frac{1}{2} = \log a.$$

For the second integral after the equals sign it follows after substitution of $(1-\frac{x}{a}) = e^{-z}$ that

$$= 2\int_0^a (1-\frac{x}{a}) \log(1-\frac{x}{a}) \, d(1-\frac{x}{a}) = 2\int_0^a e^{-z} \log e^{-z} \, d e^{-z}$$

$$= 2\log e \int_0^\infty z \cdot e^{-2z} \, dz = \frac{1}{2} \log e,$$

so that $H(X)$ becomes

$$H(X) = \log a + \frac{1}{2} \log e = \log a\sqrt{e} \quad \text{bits/sample.}$$

b. The information power of the source is defined as

$$P_{\text{H}} = \frac{1}{2\pi e} 2^{2H(X)}.$$

Substitution of $H(X)$ gives

$$P_{\text{H}} = \frac{1}{2\pi e \cdot 2^{2\log(a\sqrt{e})}} = \frac{1}{2\pi e} 2^{\log(a^2 e)} = \frac{a^2 e}{2\pi e} = \frac{a^2}{2\pi}.$$

c. The power P_f is defined as follows:

$$P_f = \text{var}(\mathbf{x}) = \int_{-\infty}^\infty \{x - E(\mathbf{x})\}^2 \, p(x) \, dx.$$

Because $E(\mathbf{x}) = 0$, this expression can be written as

$$P_f = \int_{-\infty}^\infty x^2 \, p(x) \, dx = \frac{2}{a}\int_0^a x^2(1-\frac{x}{a}) \, dx = \frac{a^2}{6}.$$

d. For the amount of information $H(Y)$ of a sample of a gaussian signal we have

$$H(Y) = \log \sigma \sqrt{2\pi e}\,.$$

Because the gaussian source must have the same power as the source being considered $\sigma = \sqrt{\dfrac{a^2}{6}}$, so that we have

$$H(Y) = \log \sqrt{\frac{2\pi\,ea^2}{6}}\,.$$

As $H(X) = \log a\sqrt{e}$ a comparison of $H(X)$ and $H(Y)$ yields

$$H(Y) - H(X) = \log \sqrt{\frac{2\pi\,ea^2}{6ea^2}} = \frac{1}{2}\log \frac{\pi}{3} = 0.034.$$

In other words, the information content $H(X)$ of the given source is smaller than the information content $H(Y)$ of the gaussian source with an equal power for both sources.

6

The continuous communication channel

6.1 The capacity of continuous communication channels

In the case of a continuous communication channel encoding and decoding have a different interpretation from that in the case of the discrete channel. For encoding for example one can now think of modulation methods such as amplitude modulation and frequency modulation or of bandwidth bounding, through the use of low pass or band filters for instance. Through this, the stochastic signal generated by the source and containing the information is transformed into a form suitable for the continuous channel.

Inevitably, noise is added to the signal in the transmission channel. At the receiving end, the transmitted signal must be reconstructed from the received signal. The capacity of a channel also plays an important role for the continuous communication model, if one wishes to determine under what requirements the maximum amount of information can be transmitted.

We will consider a continuous channel where a continuous signal $\mathbf{x}(t)$ is offered at the input and where a continuous signal $\mathbf{y}(t)$ is obtained at the output. We will again consider N samples of these signals. The probability density function of the received signal $y(t)$ for a given transmitted signal $x(t)$ can now be written as

$$q(\tilde{y}/\tilde{x}) = q(y_1,\ldots,y_N/x_1,\ldots,x_N).$$

If we assume that a sample $y_i = y(t_i)$ depends on only one sample $x_i = x(t_i)$, we can speak of a memoryless continuous channel. In that case

$$q(\tilde{y}/\tilde{x}) = \prod_{i=1}^{N} q(y_i/x_i). \tag{6.1}$$

The properties of the channel can therefore now be determined for one single pair of samples x and y which makes the analysis considerably easier. Besides the amounts of information $H(X)$ and $H(Y)$ belonging to the transmitting end and the receiving end respectively, there are a number of other quantities of importance, namely $H(X,Y)$, $H(X/Y)$ and $H(Y/X)$. Of further importance is the mutual information $I(X;Y) = I(Y;X)$, which indicates how much information x delivers about y or how much information y delivers about x.

For the two continuous stochastic variables **x** and **y**, representing samples of the signals $x(t)$ and $y(t)$ at the transmitting and receiving ends of a communication channel respectively, with joint probability density $p(x,y)$ the mutual information $I(X;Y)$ is defined as

$$I(X;Y) = \int\limits_{-\infty}^{\infty} \int\limits_{-\infty}^{\infty} p(x,y) \log \frac{p(x,y)}{p(x) \cdot q(y)} \, dxdy, \qquad (6.2)$$

which corresponds to equation (5.55). However, here we are dealing with a specific interpretation of equation (5.55) that relates **x** and **y** to the input and output of the channel, respectively.

The following relations now apply to the various information amounts:

$$I(X;Y) = H(X) + H(Y) - H(X,Y)$$

$$= H(X) - H(X/Y)$$

$$= H(Y) - H(Y/X). \qquad (6.3)$$

Because $H(Y/X) \leq H(Y)$, one can immediately see that

$$I(X;Y) \geq 0, \qquad (6.4)$$

with equality if **x** and **y** are statistically independent. In the case where **x** and **y** are identical one can also derive that $I(X;X) = H(X)$. The mutual information $I(X;Y)$ is also often denoted by the transmission rate R since it represents the amount of information transported over the channel.

Just as was the case in Section 5.3, wcorrectnuous information could only be determined under assumed constraints, constraints should be assumed on the input probability distribution of a continuous channel. Otherwise, the input of the channel could be any number on an infinite real line, which is a physically impossible situation.

The capacity C of a continuous channel is the maximum of the transmission rate R or $I(X;Y)$ that one can achieve by connecting all possible information sources, consistent with the constraints, to the channel.

$$C = \max_{p(x)} I(X;Y) = \max_{p(x)} \{H(Y) - H(Y/X)\}. \tag{6.5}$$

In general, computation of the capacity of a continuous channel is a difficult task. Only for some specific channels, e.g. channels with additive white gaussian noise with an average power limitation, is it possible to obtain an analytical expression for the capacity. In other cases numerical methods should be applied. Let us first consider the general class of the so-called additive channels, where noise is added to the transmitted signal **x** and is statistically independent of it. The value y can arise for a given value x if $n = y - x$ holds for the noise, so that we must have

$$q(y/x) = q(x + n/x) = p(n/x). \tag{6.6}$$

Because the noise is independent of the input signal, it follows that

$$q(y/x) = p(n/x) = p(n) = p(y - x). \tag{6.7}$$

For the additive channel it follows for the noise influence that

$$
\begin{aligned}
H(Y/X) &= -\int_{-\infty}^{\infty} \int_{-\infty}^{\infty} p(x,y) \log q(y/x) \, dxdy \\[2mm]
&= -\int_{-\infty}^{\infty} \int_{-\infty}^{\infty} p(x) \, q(y/x) \log q(y/x) \, dxdy \\[2mm]
&= -\int_{-\infty}^{\infty} \int_{-\infty}^{\infty} p(x) \, p(n) \log p(n) \, dxdn \\[2mm]
&= -\int_{-\infty}^{\infty} p(n) \log p(n) \, dn = H(\mathrm{N}).
\end{aligned} \tag{6.8}
$$

In that case, we find for the capacity that

$$C = \max_{p(x)} \{H(Y) - H(\mathrm{N})\} = \max_{p(x)} \{H(Y)\} - H(\mathrm{N}). \tag{6.9}$$

Thus, maximising R requires maximising $H(Y)$. But these maximisation constraints on the transmitted signals should be taken into account.

Example 6.1

The probability distribution functions with respect to **x** and **n** are given by

$$p(x) = \frac{1}{8} \quad \text{for } -4 \le x \le 4,$$
$$= 0 \quad \text{otherwise.}$$

$$p(n) = \frac{1}{2} \quad \text{for } -1 \le n \le 1,$$
$$= 0 \quad \text{otherwise.}$$

For this the joint probability distribution $p(x,y)$ can be computed, where **y** = **x** + **n**.

$$p(n) = q(y/x) = \frac{p(x,y)}{p(x)}$$

Thus, compare Figure 6.1,

$$p(x,y) = \begin{cases} \dfrac{1}{16} & \text{for } -4 \le x \le 4 \text{ and } -1 \le y - x \le 1, \\ 0 & \text{otherwise.} \end{cases}$$

The probability distributions $q(y)$ and $p(x/y)$ can be determined as follows (see Figure 6.2):

$$q(y) = \begin{cases} \displaystyle\int_{-4}^{y+1} \frac{1}{16} \, dx = \frac{1}{16}(y+5) & \text{for } -5 \le y \le -3, \\[3mm] \displaystyle\int_{y-1}^{y+1} \frac{1}{16} \, dx = \frac{1}{8} & \text{for } -3 \le y \le 3, \\[3mm] \displaystyle\int_{y-1}^{4} \frac{1}{16} \, dx = \frac{1}{16}(5-y) & \text{for } 3 \le y \le 5, \end{cases}$$

Figure 6.1. Definition domain of $p(x,y)$ of Example 6.1

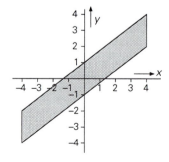

$$p(x/y) = \frac{p(x,y)}{q(y)} = \begin{cases} \dfrac{1}{y+5} & \text{for } -5 \le y \le -3, \quad -4 \le x \le y+1, \\[2mm] \dfrac{1}{2} & \text{for } -3 \le y \le 3, \quad y-1 \le x \le y+1, \\[2mm] \dfrac{1}{5-y} & \text{for } 3 \le y \le 5, \quad y-1 \le x \le 4. \end{cases}$$

With the help of the probability distribution the various measures can be computed. This is left to the reader. Here, we only give the results:

$$H(X) = 3,$$
$$H(Y/X) = H(\mathsf{N}) = 1,$$
$$H(X,Y) = 4,$$
$$H(Y) = 3 + \frac{1}{8}\log e,$$
$$H(X/Y) = 1 - \frac{1}{8}\log e.$$

Thus the information rate is equal to

$$I(X,Y) = H(Y) - H(\mathsf{N}) = 2 + \frac{1}{8}\log e.$$

The capacity could be computed by varying the input so that $H(Y)$ achieves a maximum

$$C = \lim_{p(x)} H(Y) - 1.$$

This is only possible if a constraint on the maximisation is predefined. △

In the continuous case one will generally not be able to assume that the channel is memoryless. In that case we define the transmission rate as

$$I(\tilde{X};\tilde{Y}) = \lim_{N\to\infty} \int\limits_{-\infty}^{\infty} \int\limits_{-\infty}^{\infty} p(\tilde{x},\tilde{y}) \cdot \log\frac{p(\tilde{x},\tilde{y})}{p(\tilde{x})\cdot q(\tilde{y})} \, d\tilde{x}d\tilde{y}, \tag{6.10}$$

where $\tilde{X}$ and $\tilde{Y}$ are N-dimensional stochastic vectors. The channel capacity of a channel with memory then becomes

Figure 6.2. The probability density $q(y)$ of Example 6.1.

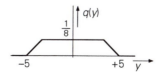

$$C = \lim_{\substack{N \to \infty}} \max_{p(\tilde{x})} I(\tilde{X};\tilde{Y}). \tag{6.11}$$

The capacity of channels with memory will be dealt with in Section 6.5.

6.2 The capacity in the case of additive gaussian white noise

The signal to be transported and the noise added in the channel are often specified by their power. We assume that the noise is additive and independent of the transmitted signal $x(t)$. This is a realistic assumption in many cases. It then follows for the (average) power P_y at the receiving end that

$$P_y = P_x + P_n, \tag{6.12}$$

where P_n is the average power of the noise. We further assume that the noise is *white*, i.e. it has a flat power density spectrum over bandwidth W ($W \to \infty$). Furthermore, we assume that its value (amplitude) at each time is characterized by a gaussian probability density function $N(0, \sigma^2)$. In that case one speaks of *gaussian white noise*. Computing the correlation function for white noise, using equation (5.39), the flat spectrum yields a delta or peak function $R_{xx}(\tau) = 0$ for $\tau \neq 0$. This implies that the gaussian white noise has no internal dependence and that the samples are stochastically independent. Therefore we can consider samples individually.

We wish to express the capacity in bits/sec and as a result will also give $H(N)$ and $H(Y)$ in bits/sec. Making use of the results of Section 5.3, it follows that the amount of information of the noise signal is

$$H(N) = \log \sigma_n \sqrt{2\pi e} \quad \text{bits/sample}$$

$$= W \log\{2\pi e \sigma_n^2\} \quad \text{bits/sec.} \tag{6.13}$$

For a signal which is limited to a certain average power this is also the maximum amount of information. This means that since the capacity of the continuous channel decreases as $H(N)$ increases, this gaussian noise corresponds with the worst case situation. Because of this, the calculation of the capacity in the case of gaussian noise will in general yield a lower bound for the actual capacity, which is quite acceptable from the point of view of a designer.

For the capacity of a continuous communication channel with additive noise we have (equation (6.9))

$$C = \max_{p(x)} \{H(Y)\} - H(\mathsf{N}) \quad \text{bits/sec.}$$

The maximum of $H(Y)$ with $\mathbf{y} = \mathbf{x} + \mathbf{n}$ will occur in the case where $\mathbf{y}$ has a gaussian distribution with power $P_y = \sigma_y^2$. Because the noise is gaussian it follows that the input signal must also be gaussian. Because now $P_y = \sigma_y^2 = \sigma_x^2 + \sigma_n^2$ it therefore follows that

$$\max_{p(x)} H(Y) = \tfrac{1}{2}\log\left\{2\pi\,\mathrm{e}(\sigma_x^2 + \sigma_n^2)\right\} \quad \text{bits/sample}$$

$$= W\log\left\{2\pi\,\mathrm{e}(\sigma_x^2 + \sigma_n^2)\right\} \quad \text{bits/sec.} \tag{6.14}$$

The channel capacity is therefore in the case of an average power constraint

$$C = W\log\left\{2\pi\,\mathrm{e}(\sigma_x^2 + \sigma_n^2)\right\} - W\log\left\{2\pi\,\mathrm{e}\,\sigma_n^2\right\}$$

$$= W\log\left\{\frac{\sigma_x^2 + \sigma_n^2}{\sigma_n^2}\right\} = W\log\left\{\frac{P_x + P_n}{P_n}\right\}$$

$$= W\log\left\{1 + \frac{P_x}{P_n}\right\} \quad \text{bits/sec.} \tag{6.15}$$

As will be seen it is not possible to transmit at a higher rate than C without introduction of errors.

From the expression found for the channel capacity it emerges that we can make a trade-off between the bandwidth W and the signal-to-noise ratio P_x/P_n, and still retain the same capacity.

6.3 Capacity bounds in the case of non-gaussian white noise

For the given derivation of the channel capacity we assumed the worst possible situation, namely a gaussian noise signal that has a maximum amount of information, while the power density spectrum is assumed to be flat so that the samples are statistically independent (memoryless channel). If these requirements are not met, a different value will be found for the capacity. The capacity will become greater if the noise is not gaussian. The analytic determination of the channel capacity will then usually be complicated, however. Numerical methods exist that can determine the capacity with a good approximation. One can also indicate upper and lower bounds for the capacity that are relatively simple to determine. One then makes use of the concept of information power P_H. We consider the situation where the noise is indeed bounded in power, but not gaussian. We further assume that the samples are statistically independent. The amount of information of the noise is (compare equation (5.73))

$$H(\mathrm{N}) = \log \sqrt{2\pi\,\mathrm{e}\,P_{H_n}} \text{ bits/sample} = W \log \left\{ 2\pi\,\mathrm{e}\,P_{H_n} \right\} \text{ bits/sec.} \tag{6.16}$$

The amount of information of the received signal, **y**, has as upper bound the value corresponding to the situation where **y** has the properties of white gaussian noise:

$$H(Y) \leq W \log \left\{ 2\pi\,\mathrm{e}(P_x + P_n) \right\}. \tag{6.17}$$

The amount of transmitted information then becomes

$$I(X;Y) \leq W \log \left\{ 2\pi\,\mathrm{e}(P_x + P_n) \right\} - W \log \left\{ 2\pi\,\mathrm{e}\,P_{H_n} \right\}, \tag{6.18}$$

so that the upper bound for the capacity is given by

$$C \leq W \log \left\{ \frac{P_x + P_n}{P_{H_n}} \right\}. \tag{6.19}$$

A lower bound on C can be obtained by considering the rate, if we make the input signal a white noise of power P_x by suitable coding and by using the result of Section 5.5 that the information power of the sum of two signals is greater than or equal to the sum of the individual information powers. Then it is the case that

$$P_x + P_{H_n} = P_{H_x} + P_{H_n} \leq P_{H_y}. \tag{6.20}$$

Using the upper bound on $H(Y)$ we find

$$H(Y) = W \log \left\{ 2\pi\,\mathrm{e}P_{H_y} \right\} \geq W \log \left\{ 2\pi\,\mathrm{e}(P_x + P_{H_n}) \right\} \tag{6.21}$$

and thus

$$I(X;Y) \geq W \log \left\{ 2\pi\,\mathrm{e}(P_x + P_{H_n}) \right\} - W \log \left\{ 2\pi\,\mathrm{e}\,P_{H_n} \right\},$$

so that

$$C \geq W \log \left\{ \frac{P_x + P_{H_n}}{P_{H_n}} \right\}. \tag{6.22}$$

Finally, by combining both of these results we find that the channel capacity is given by the inequality

$$W \log \left\{ \frac{P_x + P_{H_n}}{P_{H_n}} \right\} \; \leq \; C \; \leq \; W \log \left\{ \frac{P_x + P_n}{P_{H_n}} \right\}. \tag{6.23}$$

In the case where the noise is gaussian we have $P_{H_n} = P_n$ and the inequality given above reduces to the previously mentioned expression for the channel capacity.

Example 6.2
The input signal of a noisy channel has an average power $P_x = \sigma_x^2 = 4$.
The noise has a uniform distribution on $[-\alpha, \alpha]$. Thus the amount of information of the noise becomes

$$H(\mathsf{N}) = \log 2\alpha \text{ bits/sample} = 2W \log 2\alpha \text{ bits/sec.}$$

For the information power and the average power of the noise, it is the case that

$$P_{H_n} = \frac{1}{2\pi e} \, 2^{2H(\mathsf{N})} = \frac{2\alpha^2}{\pi e},$$

$$P_n = \int_{-\alpha}^{\alpha} n^2 \, \frac{1}{2\alpha} \, dn = \tfrac{1}{3} \, \alpha^2.$$

Substitution of the powers into equation (6.23) yields

$$W \log \left(1 + \frac{2\pi e}{\alpha^2} \right) \leq C \leq W \log \left(\tfrac{1}{6} \pi e + \frac{2\pi e}{\alpha^2} \right).$$

Figure 6.3. Capacity for gaussian white noise and bounds for non-gaussian white noise.

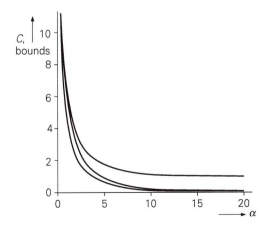

In the case where the noise was gaussian with the same average power $\frac{1}{3}\alpha^2$, the capacity would be (compare equation (6.15))

$$C = W \log \left(1 + \frac{P_x}{P_n}\right) = W \log \left(1 + \frac{12}{\alpha^2}\right).$$

See Figure 6.3.   △

6.4 Channel coding theorem

For continuous communication channels it is also the case that information can be transported with an arbitrarily small error ε provided that the amount of information at the input of the channel is less than the channel capacity. This is thus analogous to the channel coding theorem for the discrete case given in Chapter 4.

We will give a simplified proof of this channel coding theorem by proving that codes exist with which it is possible to make maximum use of the channel capacity with an arbitrarily small error ε.

Theorem 6.1 (Shannon's coding theorem for continuous channels)
It is possible to transport an amount of information $H(X)$ through a continuous white gaussian channel with capacity C with an arbitrarily small error probability ε (or equivocation) in the case where $H(X) \leq C$.   □

Proof
The continuous information source generates a stochastic signal which we will always consider over a certain time period T in this section. By assuming that the information source is ergodic, the stochastic properties of the source are equally present in the realisations of the stochastic signal with duration T.

We now consider $M(T)$ different signals each with a duration T and assume that these signals all have the same probability. It can then readily be seen that the amount of information of the source that generates these signals can be written as

$$H(X) = \lim_{T \to \infty} \frac{1}{T} \log M(T) \quad \text{bits/sec.} \tag{6.24}$$

As has been previously derived in this chapter, see equation (6.15), for the channel capacity in the case of white gaussian noise we have that

$$C = W \log \left\{1 + \frac{P_x}{P_n}\right\} \quad \text{bits/sec.}$$

We must now prove that the error probability ε can go to zero if $H(X) < C$, or in other words if

$$\lim_{T \to \infty} \frac{1}{T} \log M(T) < W \log \left\{ 1 + \frac{P_x}{P_n} \right\} . \tag{6.25}$$

For the proof of the theorem use will be made of a manner of representation where signals are represented as vectors in a high-dimensional space. According to the sampling theorem $2WT$ samples are needed in order to represent a continuous signal of duration T and with bandwidth W Hz. By now introducing a $2WT$-dimensional space it is possible to represent the signals as vectors where the values of the elements correspond to the samples and of which the tips define points in a $2WT$-dimensional space. Thus a music signal with a bandwidth of 15 kHz and a duration of 60 minutes can be represented as a point in a 10^8-dimensional space. The transmitted and received signals as well as the noise signals can be represented as vectors/points in this manner.

Denoting the $2WT$ samples by x_i, $i = 1,\ldots,2WT$, the length d of a signal vector will be equal to

$$d = \left(\sum_{i=1}^{2WT} x_i^2 \right)^{\frac{1}{2}}. \tag{6.26}$$

If the signals have an average value of zero and noting that the average power (variance) is given by

$$P = \frac{1}{2WT} \sum_{i=1}^{2WT} x_i^2, \tag{6.27}$$

we obtain

$$d = \sqrt{2WTP}. \tag{6.28}$$

Thus all signals with a power of P thus lie on a hypersphere of radius $d = \sqrt{2WTP}$ in a $2WT$-dimensional space. Here, a hypersphere can be considered as the $2WT$-dimensional generalisation of the normal three-dimensional sphere. Analogously we also speak of a hypervolume in addition to the volume of a sphere.

The volume of a $2WT$-dimensional hypersphere is proportional to the radius d to the power $2WT$:

$$V_{2WT} = \alpha_{2WT} \, d^{2WT}, \tag{6.29}$$

where α_{2WT} is given by

$$\alpha_{2WT} = \frac{\pi^{WT}}{\Gamma(WT+1)} , \tag{6.30}$$

where $\Gamma(.)$ is the gamma function.

Let us now assume that we have a signal with average power P_x, which is affected by noise with average power P_n. Each input signal can be represented as a point in the $2WT$-dimensional space, i.e. somewhere in the hypersphere with radius $d = \sqrt{2WTP_x}$.

The possible output signals are points somewhere in the hypersphere with radius $d = \sqrt{2WT(P_x+P_n)}$.

The precise position of the point of an output signal which consists of the input signal plus noise is not known. However, it should be somewhere in a sphere of radius $d = \sqrt{2WTP_n}$ centred around the point representing the transmitted signal. In Figure 6.4 this sphere is represented by a shaded circle.

Clearly, if the hyperspheres with respect to the noise do not have any overlap, there will be no interchange between transmitted and received signal. In that case the information transmission can be error-free. Let $M(T)$ be the number of transmitted signals such that the small hyperspheres with the noise are non-overlapping. The volume of the hypersphere of the output signals will be at least $M(T)$ times the volume of one of the noise hyperspheres. Thus

$$\alpha_{2WT}(\sqrt{2WT(P_x+P_n)})^{2WT} \geq M(T)\,\alpha_{2WT}(\sqrt{2WTP_n})^{2WT},$$

and

$$M(T) \leq \left(\sqrt{\frac{P_x+P_n}{P_n}}\right)^{2WT} = \left(1 + \frac{P_x}{P_n}\right)^{WT}. \tag{6.31}$$

The ratio P_x/P_n is again the well-known signal-to-noise ratio. From this it follows that

Figure 6.4. The $2WT$-dimensional hypersphere with (distorted) signals.

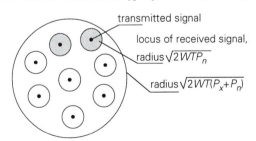

$$\lim_{T\to\infty} \frac{\log M(T)}{T} \leq W \log \left(1 + \frac{P_x}{P_n}\right). \tag{6.32}$$

Thus, then, the amount of information is less than the capacity.

That it is possible to obtain an information transmission rate very nearly equal to the capacity with the error probability ε going to zero can be seen as follows.

We fix a certain number $M(T)$ of points in this space as being related to signal, without regard for their mutual distances to avoid overlapping regions. A particular selection of $M(T)$ points constitutes a particular code for signals to be transmitted. Errors occur if a point is within the noise hypersphere of another point; then it might be wrongly identified. The probability p that such a point is within a noise hypersphere is the ratio of its volume and that of the output hypersphere:

$$p = \frac{\alpha_{2WT}\{\sqrt{2WTP_n}\}^{2WT}}{\alpha_{2WT}\{\sqrt{2WT(P_x+P_n)}\}^{2WT}}. \tag{6.33}$$

The probability that a signal is *not* within the noise hypersphere is $1 - p$. In order to have error-free reception of information $M(T) - 1$ signals should be outside of this noise hypersphere. This probability is $(1-p)^{M(T)-1}$. This probability must approach 1, hence

$$(1-p)^{M(T)-1} > 1 - \varepsilon, \tag{6.34}$$

where ε is an arbitrarily small number.

By expanding the left-hand side into a series and breaking it off after the second term, this inequality will certainly be satisfied if the following inequality holds:

$$1 - (M(T) - 1)p > 1 - \varepsilon \quad \Rightarrow \quad (M(T) - 1)p < \varepsilon$$

$$\Rightarrow \quad M(T) - 1 < \frac{\varepsilon}{p} \quad \Rightarrow \quad M(T) - 1 < \varepsilon\left\{\frac{P_x + P_n}{P_n}\right\}^{WT}. \tag{6.35}$$

A more stringent requirement is

$$M(T) < \varepsilon\left\{\frac{P_x + P_n}{P_n}\right\}^{WT}. \tag{6.36}$$

From this follows as condition

$$\lim_{T\to\infty} \frac{1}{T}\log M(T) < \lim_{T\to\infty} \frac{1}{T}\log\left[\varepsilon\left\{\frac{P_x + P_n}{P_n}\right\}^{WT}\right]$$

$$= W \log(1 + \frac{P_x}{P_n}) + \lim_{T \to \infty} \frac{\log \varepsilon}{T}. \qquad (6.37)$$

The remaining limit in the right-hand side approaches zero with increasing T. We can therefore make ε arbitrarily small, provided that we satisfy the inequality $H(X) < C$ and provided that we choose the duration T of the signals to be large.   □

Thus, there exist codes which permit transmission at a rate as close as desired to the channel capacity with an arbitrarily small error rate ε. This forms the basic idea behind Shannon's channel coding theorem.

6.5 The capacity of a gaussian channel with memory

In the foregoing sections the capacity was determined for the case where the noise is white, i.e. has a flat power density spectrum and has a gaussian probability density.

In practice the noise will often not have a flat spectrum so that the samples are not statistically independent. To determine the channel capacity in such a case (channel with memory) one should take into account the spectra of the stochastic input and noise signals $x(t)$ and $n(t)$

We assume that the input signal and noise are both bounded in power P and in bandwidth W and have a gaussian probability density. The relationship between the power and the power density spectrum is then (compare also Section 5.2)

Figure 6.5. Example of spectra of noise and input signal.

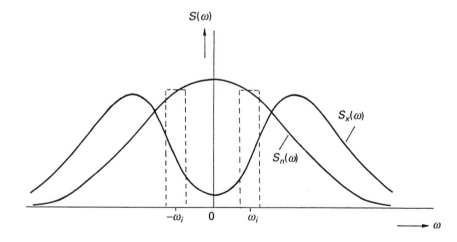

$$P_x = \frac{1}{2\pi} \int\limits_{-2\pi W}^{2\pi W} S_x(\omega)\, d\omega, \tag{6.38}$$

$$P_n = \frac{1}{2\pi} \int\limits_{-2\pi W}^{2\pi W} S_n(\omega)\, d\omega. \tag{6.39}$$

An example of two such spectra is given in Figure 6.5.

We subsequently imagine the spectrum to be divided up into regions of width $\Delta\omega$ that are so small that the spectrum can be taken to be constant in each sub-region, namely $S(\omega_i)$, and that the output signals of these fictional (very) small bandwidth filters are more or less uncorrelated. The average power per region $\Delta\omega$ is then

$$P_{x_i} = \frac{1}{2\pi} S_x(\omega_i)\, \Delta\omega,$$

$$P_{n_i} = \frac{1}{2\pi} S_n(\omega_i)\, \Delta\omega.$$

Assume the number of sub-regions to be N, then $N = 4\pi W / \Delta\omega$ holds and thus $W = N\omega_i$ where $\omega_i = \Delta\omega/4\pi$. Substitution in the expression for the channel capacity gives for the channel capacity in a sub-region

$$C_i = \omega_i \log \left\{ 1 + \frac{P_{x_i}}{P_{n_i}} \right\} = \frac{\Delta\omega}{4\pi} \log \left\{ 1 + \frac{S_x(\omega_i)}{S_n(\omega_i)} \right\}. \tag{6.40}$$

The capacity for the complete spectrum between $-2\pi W$ and $2\pi W$ is

$$C = \sum_{i=1}^{N} C_i \,,$$

that is

$$C = \frac{\Delta\omega}{4\pi} \sum_{i=1}^{N} \log \left\{ 1 + \frac{S_x(\omega_i)}{S_n(\omega_i)} \right\}. \tag{6.41}$$

If we subsequently take the limit for $\Delta\omega \to 0$ it then follows that

$$C = \frac{1}{4\pi} \int\limits_{-2\pi W}^{2\pi W} \log \left\{ 1 + \frac{S_x(\omega)}{S_n(\omega)} \right\} d\omega \quad \text{bits/sec.} \tag{6.42}$$

If the spectrum is even, then $S_x(\omega) = S_x(-\omega)$ so that we can write

$$C = \frac{1}{2\pi} \int_0^{2\pi W} \log \left\{ 1 + \frac{S_x(\omega)}{S_n(\omega)} \right\} d\omega \quad \text{bits/sec.} \tag{6.43}$$

As an illustration we examine the case where both spectra are constant. We then have

$$P_x = 2W \, S_x(\omega), \tag{6.44}$$

and

$$P_n = 2W \, S_n(\omega), \tag{6.45}$$

which reduces the channel capacity found to

$$C = W \log \left\{ 1 + \frac{P_x}{P_n} \right\} \quad \text{bits/sec,} \tag{6.46}$$

which is exactly the previously found expression for the capacity of a channel with white noise.

Thus it turns out that the channel capacity depends on the spectra of both the signal and the noise. Usually the spectrum of the noise is given, but one can still choose the spectrum of the signal. The question arises, what is the best choice for $S_x(\omega)$, or also what is the maximum value of C for a given noise spectrum $S_n(\omega)$.

The expression for the channel capacity is now rewritten as

$$C = \frac{1}{2\pi} \int_0^{2\pi W} \log \left[S_x(\omega) + S_n(\omega) \right] d\omega - \frac{1}{2\pi} \int_0^{2\pi W} \log S_n(\omega) \, d\omega. \tag{6.47}$$

The second integral is a constant for a given $S_n(\omega)$, so that only the first integral can be maximised. Since the power is bounded, to wit

$$P_x + P_n = \frac{1}{\pi} \int_0^{2\pi W} \left(S_x(\omega) + S_n(\omega) \right) d\omega, \tag{6.48}$$

the maximisation problem that must be solved is in actual fact the following one. Determine the maximum of

$$I = \int_0^{2\pi W} \log[f(\omega)] \, d\omega,$$

with as constraint

$$\int_0^{2\pi W} f(\omega) \, d\omega = \pi(P_x + P_n).$$

where $f(\omega) = S_x(\omega) + S_n(\omega)$.
The method of Lagrange gives

$$\frac{d \log[f(\omega)]}{df(\omega)} - \lambda \frac{df(\omega)}{df(\omega)} = 0,$$

or also $\quad \dfrac{\log e}{f(\omega)} - \lambda = 0,$

with which we find

$$f(\omega) = \frac{\log e}{\lambda} = \text{constant}.$$

Substitution in the constraint condition gives

$$f(\omega) = \frac{(P_x + P_n)}{2W},$$

so that

$$I = 2\pi W \log \left\{ \frac{(P_x + P_n)}{2\pi W} \right\}.$$

It follows that the capacity is maximum if

$$S_x(\omega) + S_n(\omega) = \frac{P_x + P_n}{2W} = \text{constant}. \tag{6.49}$$

In that case the channel capacity is

$$C = W \log \left\{ \frac{P_x + P_n}{2W} \right\} - \frac{1}{2\pi} \int_0^{2\pi W} \log S_n(\omega)\, d\omega. \tag{6.50}$$

Figure 6.6.

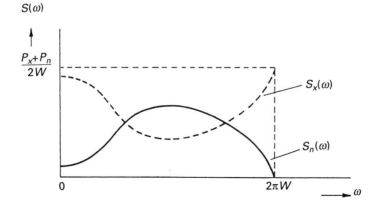

An example of this is given in Figure 6.6.

It can be seen from this that $S_x(\omega)$ must be chosen $(S_x(\omega) = \text{constant} - S_n(\omega))$ in such a way that the spectral density is large for values of ω where the noise has a small power density. The signal-to-noise ratio is as large as possible for this value of ω. Furthermore, it again follows that noise with a flat power density spectrum is the most unfavourable case, because the second term in the expression of the capacity is then maximum, causing C to become minimum.

Example 6.3

The spectra of the stochastic signals $x(t)$ and $n(t)$ are given by, respectively,

$$S_x(\omega) = \begin{cases} 4 & \text{for } 0 < |\omega| < \pi W, \\ 8 & \text{for } \pi W < |\omega| < 2\pi W \\ 0 & \text{otherwise,} \end{cases}$$

$$S_n(\omega) = \begin{cases} 1 & \text{for } 0 < |\omega| < \frac{2}{3}\pi W, \\ 2 & \text{for } \frac{2}{3}\pi W < |\omega| < \frac{4}{3}\pi W, \\ 4 & \text{for } \frac{4}{3}\pi W < |\omega| < 2\pi W, \\ 0 & \text{otherwise.} \end{cases}$$

The channel capacity becomes

$$C = \frac{1}{2\pi} \int\limits_0^{2\pi W} \log \left\{ 1 + \frac{S_x(\omega)}{S_n(\omega)} \right\} d\omega$$

$$= \frac{1}{2\pi} \left\{ \frac{2}{3}\pi W \log\left(1 + \frac{4}{1}\right) + \frac{1}{3}\pi W \log\left(1 + \frac{4}{2}\right) + \frac{1}{3}\pi W \log\left(1 + \frac{8}{2}\right) \right.$$
$$\left. + \frac{2}{3}\pi W \log\left(1 + \frac{8}{4}\right) \right\}$$

$$= \frac{1}{2} W \log 15 = 1.45W.$$

The average power equals

$$P_x = \frac{1}{\pi} \int\limits_0^{2\pi W} S_x(\omega) \, d\omega = \frac{1}{\pi}(4\pi W + 8\pi W) = 12W,$$

$$P_n = \frac{1}{\pi} \int\limits_0^{2\pi W} S_n(\omega)\, d\omega = \frac{1}{\pi} (\tfrac{2}{3}\pi W + \tfrac{4}{3}\pi W + \tfrac{8}{3}\pi W) = \tfrac{14}{3}\, W.$$

Thus the signal-to-noise ratio yields $12W/(\tfrac{14}{3}\, W) = \frac{18}{7}$.
The capacity is maximum if

$$S_x(\omega) + S_n(\omega) = \frac{P_x + P_n}{2W} = \frac{\tfrac{14}{3} + \tfrac{36}{3}}{2} = \frac{25}{3}.$$

The maximal channel capacity then is

$$C = W \log \left\{ \frac{P_x + P_n}{2W} \right\} = W \log \tfrac{25}{3} = 2.06W. \qquad \triangle$$

6.6 Exercises

6.1. The input signal of a continuous communication channel is denoted by *x,* and the output signal by *y.* For this channel the joint probability density $p(x,y)$ of both signals is uniform in the region G given by $0 < x < 1$, $0 < y < x$. and $p(x,y) = 0$ outside of this region G.
a. Calculate the amount of information of the source.
b. Calculate the amount of information at the receiving end.
c. Calculate the equivocation.
d. Calculate the amount of information in the joint occurrence of x and y.
e. Calculate the amount of transmitted information.
f. Verify the relationship $H(X,Y) \le H(X) + H(Y)$ for this channel.

6.2. An information source delivers a signal with a negative-exponential probability density of the amplitude:

$$p(x) = e^{-x}, \quad 0 < x < \infty.$$

This source is connected to a channel that is distorted by independent, additive noise with uniform probability density of the amplitude between the amplitude values 0 and α. The probability density is zero outside of this region. Calculate the joint amount of information with respect to input and output.

6.3. *a.* Give an expression for the channel capacity of a continuous channel that is disturbed by ideal noise. This noise is additive and independent of the signal $x(t)$ (which is bounded in power).
b. Sketch the channel capacity as a function of the signal-to-noise ratio P_x/P_n for a constant bandwidth.

c. Calculate the upper bound for the channel capacity if the bandwidth increases limitlessly and if it is given that the mean power of the noise per unit of bandwidth is constant and equal to P_n^o.

d. Sketch the channel capacity as a function of the bandwidth under the requirement given in c.

e. If the channel capacity of the aforementioned channel is 5.6×10^4 bits/sec and one can choose between two bandwidths for the signal to be transmitted, 7 kHz and 8 kHz, what signal-to-noise ratios should then be realised for both cases?

f. Can you say anything about the preference that one may have for one signal above the other?

6.4. A memoryless information source delivers a signal $x(t)$ with a gaussian probability density $N(0,1)$. This source is connected to a channel that is disturbed by independent additive noise with the following probability density of the amplitude:

$$p(n) = n^2 \quad \text{for } |n| \le a,$$

$$= 0 \quad \text{for } |n| > a.$$

The power density spectrum of both the signal and the noise is bounded between $-W$ and $+W$. The sampling frequency is such that the requirements of the sampling theorem are met.

a. Calculate the amount of information of the noise per sample.

b. Calculate the information power of the noise per sample.

c. Calculate the amount of information of the noise per second.

d. Calculate the upper bound for the amount of information per second in the received signal $y(t)$ at the output of the channel.

e. Give the upper bound for the capacity of this channel.

6.5. One wishes to transmit a signal $x(t)$ over a noisy channel with memory.

a. Give an expression for the channel capacity in terms of the spectra of the signal and noise.

b. Consequently, give a relation between the signal spectrum and the noise spectrum, for which the capacity is maximum for a given noise spectrum $S_n(\omega)$.

c. If the noise spectrum $S_n(\omega)$ is given by

$$S_n(\omega) = N \qquad \text{for} \qquad 0 < |\omega| < 2\pi W_1,$$

$$= 2N \qquad \text{for} \quad 2\pi W_1 < |\omega| < 2\pi W_2,$$

$$= 0 \qquad \text{elsewhere,}$$

calculate the maximum channel capacity, in the case where it is also given that the signal-to-noise ratio is equal to 2 and that $W_2 = 2W_1$.

d. Sketch the power density spectra of the signal $S_x(\omega)$ and noise $S_n(\omega)$ in one figure. What can you say about the choice of $S_x(\omega)$?

6.7 Solutions

6.1. *a.* See Figure 6.7.

The amplitude of the probability density must be determined first. This can be done using

$$\int_{-\infty}^{\infty} \int_{-\infty}^{\infty} p(x,y)\, dxdy = 1.$$

Now suppose that $p(x,y) = k$, then

$$\int_{0}^{1}\int_{0}^{x} k\, dx\, dy = k\int_{0}^{1} dx \int_{0}^{x} dy = k\int_{0}^{1} x\, dx = \tfrac{1}{2}k = 1;$$

therefore $p(x,y) = k = 2$.

Further, $p(x)$ must be calculated from the joint probability density by integrating over y:

Figure 6.7. Definition domain of $p(x,y)$ of Exercise 6.1.

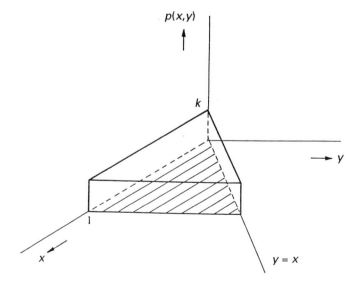

$$p(x) = \int\limits_{0}^{x} p(x,y) \, dy = 2y \Big|_{0}^{x} = 2x.$$

It then follows that

$$H(X) = -\int\limits_{-\infty}^{\infty} p(x) \log p(x) \, dx$$

$$= -\int\limits_{0}^{1} 2x \log 2x \, dx$$

$$= -\frac{1}{2}\int\limits_{0}^{1} 2x \log 2x \, d(2x)$$

$$= -\log 2 + \frac{1}{2} \log e = \log \frac{\sqrt{e}}{2} \text{ bits.}$$

b. $H(Y)$ is calculated in the same manner. We find

$$q(y) = \int\limits_{y}^{1} p(x,y) \, dx = 2x \Big|_{y}^{1} = 2(1-y),$$

and

$$H(Y) = -\int\limits_{0}^{1} 2(1-y) \log 2(1-y) \, dy = -\frac{1}{2}\int\limits_{0}^{1} 2z \log 2z \, d\,(2z).$$

This is the same expression as that for $H(X)$ in a, so that it easily follows that

$$H(Y) = H(X) = \log \frac{\sqrt{e}}{2} \text{ bits.}$$

c. $p(x/y)$ must be determined first. We have

$$p(x/y) = \frac{p(x,y)}{q(y)} = \frac{2}{2(1-y)} = \frac{1}{1-y}.$$

Now

$$H(X/Y) = -\int\limits_{-\infty}^{\infty}\int\limits_{-\infty}^{\infty} p(x,y) \log q(x/y) \, dxdy$$

$$= -\int\limits_{0}^{1}\int\limits_{0}^{x} 2 \log \frac{1}{1-y} \, dxdy = 2\int\limits_{0}^{1}\int\limits_{0}^{x} \log(1-y) \, dxdy.$$

After partial integration and substitution of the limits this becomes:

$$H(X/Y) = -\log \sqrt{e} \quad \text{bits.}$$

d. This becomes

$$H(X,Y) = -\int\limits_{-\infty}^{\infty} \int\limits_{-\infty}^{\infty} p(x,y) \cdot \log p(x,y) \, dxdy = -2\int\limits_{0}^{1} \int\limits_{0}^{x} dxdy = -1 \quad \text{bit.}$$

e. Use can be made of

$$R = H(X) - H(X/Y).$$

This gives

$$R = \log \frac{\sqrt{e}}{2} + \log \sqrt{e} = \log \frac{e}{2} \quad \text{bits/sec.}$$

f. The relation is

$$H(X,Y) \leq H(X) + H(Y).$$

Filling in the values found leads to

$$-1 \leq \log \frac{\sqrt{e}}{2} + \log \frac{\sqrt{e}}{2} \quad \Leftarrow \quad -1 \leq \log \frac{e}{4}.$$

The relation is thus indeed satisfied.

6.2. The following amplitude probability densities are given for signal and noise:

$$p(x) = e^{-x} \qquad \text{for } 0 \leq x < \infty,$$

$$p(n) = \frac{1}{\alpha} \qquad \text{for } 0 \leq n \leq \alpha \quad (\text{because } \int\limits_{0}^{\alpha} p(n) \, dn = 1).$$

For the joint amount of information we have

$$H(X,Y) = H(X) + H(Y/X) = H(X) + H(N),$$

because the noise is independent of the transmitted signal *x*. It can now be calculated that

$$H(X) = -\int\limits_{0}^{\infty} e^{-x} \log e^{-x} \, dx = \log e \quad \text{bits}$$

and that

$$H(\mathrm{N}) = -\int_0^\alpha \frac{1}{\alpha} \, \log\frac{1}{\alpha} \, dn = \log\alpha \quad \text{bits,}$$

so that, finally,

$$H(X,Y) = \log e + \log\alpha = \log\alpha e \quad \text{bits.}$$

6.3. *a.* This channel capacity is given by

$$C = W \log\left\{1 + \frac{P_x}{P_n}\right\}.$$

b. This capacity is a logarithmic function of the signal-to-noise ratio (see Figure 6.8).

c, The power of the noise is now

$$P_n = W P_n^o,$$

where W is the bandwidth of the noise. The channel capacity then becomes

$$C = W \log\left\{1 + \frac{P_x}{WP_n^o}\right\} = \log e \cdot \ln\left\{1 + \frac{P_x}{WP_n^o}\right\}^W$$

$$= \log e \, \frac{P_x}{P_n^o} \, \ln\left\{1 + \frac{P_x}{WP_n^o}\right\}^{WP_n^o/P_x}$$

Figure 6.8. Capacity as function of signal-to-noise ratio.

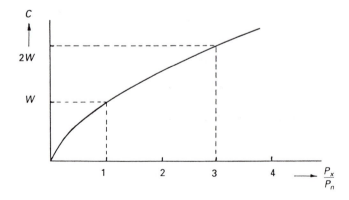

so that

$$\lim_{W \to \infty} C = \frac{P_x}{P_n^{\,o}} \log e.$$

d. Depicted graphically, $C = f(W)$ gives Figure 6.9.

e.
$$C = W \log\left\{1 + \frac{P_x}{P_n}\right\} \Rightarrow$$

$$\frac{P_x}{P_n} = 2^{C/W} - 1 = 2^8 - 1 = 255,$$

in the one case, and

$$\frac{P_x}{P_n} = 2^7 - 1 = 127,$$

in the other case.

f. There is no preference to be given to one or the other case on information theoretical grounds, because only the capacity is of importance. A possible preference must be based on other criteria, such as the fact that a twice as large signal-to-noise ratio is sometimes more difficult to realise than a 14% larger bandwidth.

6.4. *a.* See Figure 6.10.
It is possible to calculate a with the help of

$$\int_{-a}^{a} p(n) \, dn = \int_{-a}^{a} n^2 \, dn = \frac{1}{3} n^3 \Big|_{-a}^{a} = \frac{2}{3} a^3 = 1.$$

Figure 6.9. Capacity as function of bandwidth.

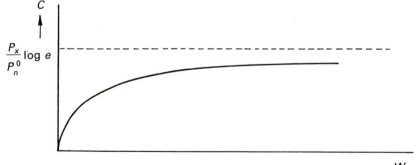

This gives $a = \sqrt[3]{\frac{3}{2}}$.

$H(\mathrm{N})$ can now be written as

$$H(\mathrm{N}) = H(Y/X) = -\int_{-\infty}^{\infty} p(n) \log p(n) \, dn$$

$$= -\int_{-a}^{a} n^2 \log n^2 \, dn = -\frac{2}{3}\int_{-a}^{a} \log n \, dn^3$$

$$= -\frac{2}{3} n^3 \log |n| \Big|_{-a}^{a} + \frac{2}{3}\int_{-a}^{a} n^3 \, d\log n$$

$$= \left[-\frac{2}{3} n^3 \log |n| + \frac{2}{3}\log e \, \frac{1}{3} n^3 \right] \Big|_{-a}^{a}$$

$$= -\frac{4}{3} a^3 \log a + \frac{4}{9} a^3 \log e = \frac{2}{3}\log \frac{2e}{3} = 0.57 \quad \text{bit/sample.}$$

b. The information power of the noise is defined as

$$P_{H_n} = \frac{1}{2\pi e} \, 2^{2H(\mathrm{N})},$$

so that

$$P_{H_n} = \frac{1}{2\pi e} \, 2^{2 \cdot 0.57} = \frac{1}{2\pi e} \, 2^{1.14} = 0.129.$$

c. The amount of information of the noise per second is

$$H(\mathrm{N})_{\text{sec}} = 2W \, H(\mathrm{N})_{\text{sample}} = 2W \, 0.57 = 1.14 \, W \quad \text{bits/sec}$$

Figure 6.10. Probability density of noise of Exercise 6.4.

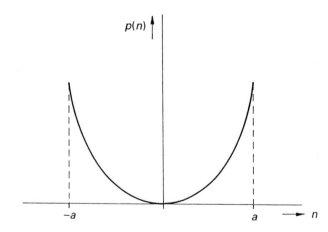

or

$$H(N)_{sec} = W \log (2\pi eP_{H_n}) = 1.14 \ W \quad \text{bits/sec.}$$

d. The amount of information of the received signal $y(t)$ has the following value as upper bound, that is if it has the properties of ideal noise:

$$H(Y) = W \log 2\pi e(P_x + P_n),$$

with

$$P_x = \sigma_x^2 = 1,$$

and

$$P_n = \sigma_n^2 = E(\mathbf{n}^2) = \int\limits_{-a}^{a} n^2 p(n) \ dn = \int\limits_{-a}^{a} n^2 n^2 \ dn = \tfrac{1}{5}n^5 \ \Big|_{-a}^{a} = \tfrac{2}{5} a^5$$

$$= \tfrac{2}{5} \left(\tfrac{3}{2}\right)^{5/3},$$

so that

$$H(Y) = W \log 2\pi e\left(1 + \tfrac{2}{5} \left(\tfrac{3}{2}\right)^{5/3}\right).$$

e. The amount of transmitted information then becomes

$$R \leq H(Y) - H(Y/X) \leq H(Y) - H(N)$$
$$\leq W \log 2\pi e(P_x + P_n) - W \log 2\pi e \ P_{H_n},$$

so that the upper bound for the capacity is given by

$$C \leq W \log \left\{ \frac{P_x + P_n}{P_{H_n}} \right\} = W \log \left\{ \frac{1 + \tfrac{2}{5}\left(\tfrac{3}{2}\right)^{5/3}}{0.129} \right\}.$$

6.5. *a.* It has been derived for the capacity of a channel with memory that:

$$C = \frac{1}{4\pi} \int\limits_{-2\pi W}^{2\pi W} \log \left\{ 1 + \frac{S_x(\omega)}{S_n(\omega)} \right\} d\omega \quad \text{bits/sec.}$$

The spectra of signal and noise are even, i.e. symmetric with regard to $\omega = 0$, so that we can write

$$C = \frac{1}{2\pi} \int\limits_{0}^{2\pi W} \log \left\{ 1 + \frac{S_x(\omega)}{S_n(\omega)} \right\} d\omega \quad \text{bits/sec.}$$

b. The expression for the channel capacity can be written as follows:

$$C = \frac{1}{2\pi} \int_0^{2\pi W} \log\left[S_x(\omega) + S_n(\omega)\right] d\omega - \frac{1}{2\pi} \int_0^{2\pi W} \log S_n(\omega)\, d\omega.$$

Since the spectrum of the noise is given, maximisation of C means maximisation of the first integral. This yields the following relation between $S_n(\omega)$ and $S_x(\omega)$:

$$S_x(\omega) + S_n(\omega) = \frac{P_x + P_n}{2W} = \text{constant}.$$

c. The expression for the maximum channel capacity becomes, with the help of the result of b (see also Figure 6.11),

$$C = W \log\left\{\frac{P_x + P_n}{2W}\right\} - \frac{1}{2\pi} \int_0^{2\pi W} \log S_n(\omega)\, d\omega.$$

For the noise power P_n we find

$$P_n = \frac{1}{\pi} \int_0^{2\pi W_1} N\, d\omega + \frac{1}{\pi} \int_{2\pi W_1}^{2\pi W_2} 2N\, d\omega = \frac{N}{\pi}\, \omega\Big|_0^{2\pi W_1} + \frac{2N}{\pi}\, \omega\Big|_{2\pi W_1}^{2\pi W_2}$$

$$= 2NW_1 + 4N(W_2 - W_1).$$

Using the fact that $W_2 = 2W_1$, P_n becomes

$$P_n = NW_2 + 2NW_2 = 3NW_2.$$

It is further given that the signal-to-noise ratio $P_x/P_n = 2$, so that

$$P_x = 2P_n = 6NW_2.$$

Substitution in the formula for the channel capacity C gives

Figure 6.11. Spectrum of noise of Exercise 6.5.

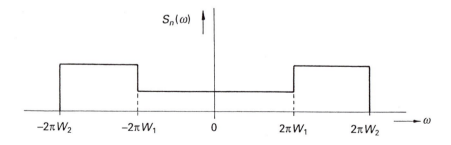

$$C = W_2 \log\left\{\frac{6NW_2 + 3NW_2}{2W_2}\right\} - \frac{1}{2\pi}\int\limits_{0}^{2\pi W_1} \log N \, d\omega - \frac{1}{2\pi}\int\limits_{2\pi W_1}^{2\pi W_2} \log 2N \, d\omega$$

$$= W_2 \log 4.5N - W_1 \log N - \log 2N \cdot [W_2 - W_1]$$

$$= W_2 \log 4.5N - \tfrac{1}{2}W_2 \log N - \tfrac{1}{2}W_2 \log 2N$$

$$= W_2 \log \frac{4.5N}{\sqrt{N}\sqrt{2N}} = W_2 \log \frac{4.5}{\sqrt{2}} = W_2 \log 3.15$$

$$= 1.66 \, W_2 \text{ bits/sec.}$$

d. See Figure 6.12.

The power density spectrum $S_x(\omega)$ is large for those values of ω for which the noise has a small power density spectrum $S_n(\omega)$.

Figure 6.12.

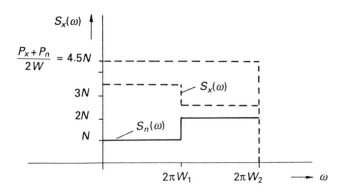

7

Rate distortion theory

7.1 The discrete rate distortion function

The theory discussed thus far exhibits two essential properties. The first is the possibility of error-free transmission if the amount of information to be transmitted is less than the channel capacity C, as was proven by the channel coding theorem (Theorem 4.5). For this to be true, the messages must be able to consist of an infinitely large number of symbols. The second property is that we assume a given channel and adapt the source to this channel. Thus the channel capacity is defined as the maximum amount of transmitted information, where the maximum is found by connecting all possible sources to the channel.

These two properties are not always realistic in practice. In practice there will be an error probability which may be small, but certainly not equal to zero. For a continuous information source a completely correct reconstruction at the receiver will even be impossible. There will always be a certain amount of distortion, however small it may be. In addition, the source will often have to be regarded as given in practice (think of a speech signal or a television signal) and the channel will have to be adapted to the source. This is done by taking a channel with just enough capacity to achieve a good transmission.

It is natural to see distortion as something which is unavoidable, but it is quite possible that distortion is introduced on purpose in some cases. With data compression a sequence of source output symbols is transformed into a sequence of symbols from a reproducing alphabet in such a way that the entropy of the new sequence is less than that of the original sequence, but at the cost of some distortion. There should, in fact, be a correspondence between each input sequence and the output sequence; the same input sequence will lead to the same output sequence. Data compression is a deterministic process. However, at the level of single symbols the relation between a source symbol u_j and the symbol $\hat{u}_k$ into which it is encoded is not one-to-one, but is effected by the transition probability $q(\hat{u}_k/u_j)$, even though at the sequence/block level the relation is deterministic.

From this chapter it may be concluded that in a well designed data compression code, the reproducing symbols of the code words should be such that the transition matrix of the conditional probabilities of reproducing symbols given the source symbols minimises the mutual information between source words and code words, subject to the constraint on average distortion.

It is characteristic for the rate distortion theory that a combination of source and reproduction is considered. A distortion measure is introduced for this combination which indicates the performance which can be expected for the reproduction of the signals or symbols produced by the source.

Before introducing a quantitative measure for distortion the information source will first be considered in more detail. For this source we assume a stationary discrete source. We further assume that the source is memoryless and so does not possess any Markov properties.

It is assumed that the source symbols are from a finite alphabet containing n symbols given by $\{u_1,...,u_j,...,u_n\}$. The source output is reproduced in terms of the so-called *reproducing alphabet* or *code word alphabet* containing m symbols $\{\hat{u}_1,...,\hat{u}_k,...,\hat{u}_m\}$.

The distortion measure to be defined must confer a certain weight on every combination of a source symbol u_j and a reproduced symbol $\hat{u}_k$. To that end we introduce a non-negative function, the *symbol distortion measure* $\rho(j,k) = \rho(u_j, \hat{u}_k)$; the distortion associated with producing the symbol $\hat{u}_k$ whereas the source symbol u_j was generated. Further, we assume that this distortion is independent of the location (or time) at which the symbols appear.

The distortion is indicated through the *distortion matrix* which depicts the distortion between the source and destination symbols. For a system with two source symbols u_1 and u_2 and two destination symbols $\hat{u}_1$ and $\hat{u}_2$ we can choose for example as distortion

$$\rho(u_j, \hat{u}_k) = 0 \qquad \text{for } j = k,$$
$$= 1 \qquad \text{for } j \neq k.$$

This gives the diagram of Figure 7.1.
Through this choice of the distortion matrix, a relation has been created with the errors that can occur between source and reproduction. A correct reproduction gives no distortion, an error introduces a distortion of one.

The distortion matrix does not have to be symmetrical. Thus the combination $u_1, \hat{u}_2$ can possess a distortion $\rho(1,2)$ which is much greater than $\rho(2,1)$. Since the source symbols have a certain probability of occurring, the next step is to determine the average distortion that occurs if a long sequence of symbols is generated. The probability of reproducing symbol $\hat{u}_k$ if u_j has been transmitted will be denoted by $q(\hat{u}_k/u_j) = q(k/j)$. If $p(j)$ is the probability of source symbol u_j occurring, then the probability of the joint occurrence of both u_j and $\hat{u}_k$ is given by

$$p(j,k) = p(j)\, q(k/j).$$

We now find the *average distortion* $d(Q)$ by multiplying the distortion of each combination $u_j, \hat{u}_k$ by the probability of their occurrence.

$$d(Q) = \sum_j \sum_k p(j)\, q(k/j)\, \rho(j,k). \tag{7.1}$$

If it is given that the distortion allowed is equal to D, then it must hold that

$$d(Q) \le D. \tag{7.2}$$

In the expression of the average distortion $d(Q)$, the source and thus $p(j)$ and the distortion $\rho(j,k)$ are given. The transition probabilities $q(k/j)$, which represent the transfer between source and reproduction, may be freely chosen, so that the average distortion is a function of these transition probabilities, that is of the matrix Q of transition or conditional probabilities. We now introduce the set Q_D which consists of the matrices Q of the transitional probabilities $q(k/j)$ for which $d(Q) \le D$.

$$Q_D = \{Q : d(Q) \le D\}. \tag{7.3}$$

Figure 7.1. Example of distortions and their matrix.

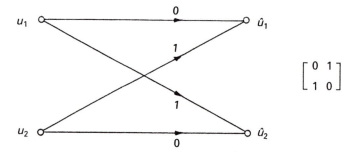

Example 7.1
For the source and reproducing alphabets it is given that $n = 2$ and $k = 3$, respectively. The probabilities of the source symbols are

$$p(u_1) = p(u_2) = \frac{1}{2}.$$

The distortion matrix is given by

$$Q = \begin{bmatrix} 0 & 1 & 3 \\ 3 & 1 & 0 \end{bmatrix},$$

which means that there is no distortion if source symbol u_1 is encoded into symbol $\hat{u}_1$ and u_2 into $\hat{u}_3$, otherwise there is distortion.
Let the permissible average symbol distortion be less than or equal to $D = 0.45$. Let two transition matrices Q_1 and Q_2 be given by

$$Q_1 = \begin{bmatrix} 0.7 & 0.2 & 0.1 \\ 0.1 & 0.2 & 0.7 \end{bmatrix}, \quad Q_2 = \begin{bmatrix} 0.8 & 0.1 & 0.1 \\ 0.1 & 0.1 & 0.8 \end{bmatrix}.$$

For both matrices, we may confirm whether or not they give an average symbol distortion satisfying the given constraint.
We find

$$d(Q_1) = \sum_{j=1}^{2} \sum_{i=1}^{2} p(j) \, q(k/j) \, \rho(j,k)$$

$$= \frac{1}{2} \cdot 0.7 \cdot 8 + \frac{1}{2} \cdot 0.2 \cdot 1 + \frac{1}{2} \cdot 0.1 \cdot 3 + \frac{1}{2} \cdot 0.1 \cdot 3 + \frac{1}{2} \cdot 0.2 \cdot 1 + \frac{1}{2} \cdot 0.7 \cdot 0$$

$$= \frac{1}{2}.$$

Similarly we find

Figure 7.2. Distortions and transition probability of Example 7.1.

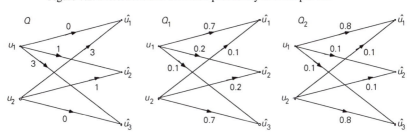

$$d(Q_2) = \frac{2}{5}.$$

Clearly, only Q_2 satisfies the constraint

$$d(Q_2) \leq D = 0.45.$$

Evidently a distortion of less than D is possible with matrix Q_2. This matrix therefore belongs to Q_D, as opposed to matrix Q_1.

The form of the distortion matrix makes it possible to see $d(Q)$ as the probability that an error occurs during transfer. D is thus the permissible error probability for reproduction and Q_D is the set of matrices of which the average error probability is smaller than or at most equal to the permissible error probability. $\triangle$

By introducing distortion when reproducing symbols, a measure for the quality of reproduction of the source is obtained. In order to come to a definition of the rate distortion function, we consider the mutual information between source and destination (reproduction). In the present case we have

$$I(U;\hat{U}) = H(U) - H(U/\hat{U})$$

$$= H(\hat{U}) - H(\hat{U}/U)$$

$$= \sum_j \sum_k p(j) \, q(k/j) \, \log \frac{p(j,k)}{p(j) \cdot q(k)}. \tag{7.4}$$

This mutual information is a measure for the correlation between U and $\hat{U}$.

In the rate distortion theory the source, that is $p(j)$, is given so that the mutual information will be regarded as a function of the matrix Q of the transition probabilities $q(k/j)$:

$$I(Q) = \sum_j \sum_k p(j) \, q(k/j) \, \log \frac{q(k/j)}{q(k)}. \tag{7.5}$$

Each matrix Q can yield a different value of $I(Q)$. We now seek the matrix for which $I(Q)$ is minimum and define the rate of the source for a permissible distortion D or simply the *rate distortion function* $R(D)$ as follows:

$$R(D) = \min_{Q \in Q_D} I(Q) \text{ bits/symbol.} \tag{7.6}$$

This function is the minimum value of the mutual information, which is achieved by varying over all the matrices Q. The matrices, however, must satisfy the prerequisite that the distortion $d(Q)$ which arises from this matrix is smaller than the permissible distortion D, in other words that $Q \in Q_D$. The source and distortion are assumed to be given.

The justification for this definition resides within some interesting theorems which will be presented in Section 7.4. It follows that the rate distortion function $R(D)$ can be regarded as the minimum rate needed by any block code for data compression, such that the average symbol distortion is less than or equal to D. Beyond that rate, no compression codes exist with distortion less than or equal to D. In Sections 7.2 and 7.3 the properties of the $R(D)$-function will first be studied in detail.

Example 7.2
In Example 7.1 two transition matrices Q_1 and Q_2 were considered. For Q_2 only, $Q_2 \in Q_D$ held, with $D = 0.45$.
For $I(Q)$ it is the case that

$$I(Q) = \sum_{j=1}^{2} \sum_{k=1}^{3} p(j)q(k/j) \log \frac{q(k/j)}{q(k)} \; .$$

Since

$$q(k) = \sum_{j=1}^{2} p(j)q(k/j),$$

we easily find

$$q(1) = \tfrac{2}{5}, \quad q(2) = \tfrac{1}{5}, \quad q(3) = \tfrac{1}{5} \; .$$

This results in

$$I(Q_2) = 2\{\tfrac{1}{2} \cdot \tfrac{7}{10} \log \tfrac{7/10}{2/5} + \tfrac{1}{2} \cdot \tfrac{2}{10} \log \tfrac{2/10}{1/5} + \tfrac{1}{2} \cdot \tfrac{1}{10} \log \tfrac{1/10}{2/5}\}$$

$$= -\tfrac{8}{5} + \tfrac{7}{10} \log 7 \approx 0.37.$$

In the following it will be considered whether this $I(Q_2)$ is the minimum such that $R(D) = I(Q_2)$. △

7.2 Properties of the $R(D)$ function
In treating the properties of the $R(D)$ function a discrete memoryless information source will be assumed, just as previously has been the case. We assume that the source has n different symbols.

Firstly, we have to see how $R(D)$ proceeds as a function of D. For such a source, the rate distortion function typically has the form shown in Figure 7.3.

The range of $R(D)$ is given by

$$0 \le R(D) \le H(U). \tag{7.7}$$

This can be shown as follows. From

$$0 \le H(U/\hat{U}) \le H(U),$$

and

$$0 \le H(U) \le \log n,$$

it follows with

$$I(U;\hat{U}) = H(U) - H(U/\hat{U})$$

that

$$0 \le I(U;\hat{U}) \le H(U) \le \log n. \tag{7.8}$$

Since $R(D)$ is the minimum of $I(U;\hat{U})$ we have $R(D) \le I(U;\hat{U})$, whereby we have found the range of $R(D)$.

Next we shall determine what value the permissible distortion D may take on. The average distortion $d(Q)$ is defined as

$$d(Q) = \sum_{j} \sum_{k} p(j)\, q(k/j)\, \rho(j,k) \le D. \tag{7.9}$$

The smallest possible average distortion $D_{\min}$ is obtained if for each source symbol u_j we seek the destination symbol $\hat{u}_k$, where $\rho(j,k)$ is minimum, and

Figure 7.3. The rate distortion function as function of D.

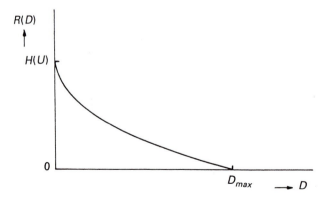

subsequently set this transitional probability $q(k/j) = 1$ and thus set the remaining probabilities equal to zero.

We define

$$\rho(j) = \min_{k} \rho(j,k), \tag{7.10}$$

hence

$$D_{\min} = \sum_{j=1}^{n} p(j)\,\rho(j). \tag{7.11}$$

Without loss of generality we will assume that $D_{\min} = 0$.

If this is not the case we modify the distortion $\rho(j,k)$ in such a manner that $D_{\min}$ becomes equal to zero again. This means that for every symbol u_j there must be a symbol $\hat{u}_k$ with $\rho(j,k) = 0$.

Since for the transitional probabilities we have

$$q(k/j) = 1 \text{ if } u_j \text{ and } \hat{u}_k \text{ correspond to each other (i.e. } q(j,k) = 0),$$

$$= 0 \text{ otherwise,}$$

the equivocation will be equal to 0, so that $I(Q) = H(U)$. Q no longer appears here so that it also directly follows that

$$R(0) = I(Q) = H(U). \tag{7.12}$$

In order to achieve zero distortion all of the information of the source must therefore be reproduced at the destination.

The maximum possible average distortion $D_{\max}$ can be determined as follows. It can be shown that the rate distortion function is monotone decreasing for increasing distortion D. The maximum D then occurs for minimal $R(D)$, thus for $R(D) = 0$. This means that the destination receives no information from the source. We denote the smallest possible distortion that is then still possible by $D_{\max}$; this value is achieved by constantly choosing that value $\hat{u}_k$ for which the average distortion between an arbitrary source symbol and the symbol $\hat{u}_k$ in question is smallest. Every other choice would lead to a larger average distortion. The mutual information $I(Q)$ is zero if $q(k/j) = q(k)$. On the one hand this follows from the definition of $I(Q)$ and it can be explained on the other hand by realising that $q(k/j) = q(k)$ means that the occurrence of u_j does not have any effect on the reproduction of $\hat{u}_k$, so that no information is reproduced. In that case, the average distortion $d(Q)$ becomes

$$d(Q) = \sum_k q(k) \sum_j p(j) \, \rho(j,k). \tag{7.13}$$

The minimum value of $d(Q)$ is obtained by setting $q(k) = 1$ for that value of k for which $\sum_j p(j) \rho(j,k)$ is smallest. This gives

$$D_{\max} = \min_k \sum_j p(j) \, \rho(j,k).$$

Example 7.3
Example 7.1 is again reconsidered, now in order to find $D_{\min}$ and $D_{\max}$ and the corresponding values of the rate distortion function.
(*i*) $D_{\min} = 0$ appears if

$$Q = \begin{bmatrix} 1 & 0 & 0 \\ 0 & 0 & 1 \end{bmatrix},$$

which corresponds to $R(0) = I(Q) = H(U) = 1$.

(*ii*) $$D_{\max} = \min_k \sum_{j=1}^{2} p(j) \, \rho(j,k)$$

$$= \tfrac{1}{2} \min_k \left(\rho(1,k) + \rho(2,k) \right)$$

$$= \tfrac{1}{2} \min \{ 0 + 3, \, 1 + 1, \, 3 + 0 \} = 1.$$

The corresponding value of $R(D)$ satisfies

$$R(D_{\max}) = R(1) = 0.$$

The corresponding transition matrix is

$$Q = \begin{bmatrix} 0 & 1 & 0 \\ 0 & 1 & 0 \end{bmatrix}. \qquad\qquad \triangle$$

The determination of the rate distortion function can be regarded as a problem from the calculus of variations. We formulate this as follows: determine the minimum of the mutual information

$$I(Q) = \sum_j \sum_k p(j) \, q(k/j) \log \frac{q(k/j)}{q(k)} , \tag{7.14}$$

as a function of $q(k/j)$ with the following constraints:

(i) $\qquad q(k/j) \geq 0,$

(ii) $\qquad \displaystyle\sum_k q(k/j) = 1,$ $\hfill (7.15)$

(iii) $\qquad \displaystyle\sum_j \sum_k p(j)\, q(k/j)\, \rho(j,k) = D.$ $\hfill (7.16)$

If we ignore constraint (i) a solution is possible by differentiating the function with respect to $q(k,j)$ and equating the result to zero. The derivation is not given here, but yields the following expression for the transition probabilities (Lagrange's method):

$$q(k/j) = \frac{q(k)\, e^{s\rho(j,k)}}{\displaystyle\sum_k q(k)\, e^{s\rho(j,k)}}. \qquad (7.17)$$

By now introducing

$$\lambda(j) = \frac{1}{\displaystyle\sum_k q(k)\, e^{s\rho(j,k)}}, \qquad (7.18)$$

it follows that

$$q(k/j) = \lambda(j)\, q(k)\, e^{s\rho(j,k)}. \qquad (7.19)$$

This yields a set of equations for each j and k, where $q(k/j)$ is expressed in terms of $q(k)$. It therefore now remains to determine the probabilities $q(k)$. In general,

$$q(k) = \sum_j p(j)\, q(k/j). \qquad (7.20)$$

This gives, if we also divide by $q(k)$,

$$\sum_j \lambda(j)\, p(j)\, e^{s\rho(j,k)} = 1, \qquad (7.21)$$

or

$$\sum_j \frac{p(j)\, e^{s\rho(j,k)}}{\displaystyle\sum_k q(k)\, e^{s\rho(j,k)}} = 1. \qquad (7.22)$$

It must still be verified if constraint (i) is satisfied. In this manner n equations are obtained for the n probabilities $q(k)$, which may now be solved. After this the transition matrix Q can be determined. With the equations so obtained it is now possible to obtain an expression for $R(D)$ and (D). Substitution of $q(k/j)$ in $d(Q)$ gives

$$D = \sum_j \sum_k \lambda(j)\, p(j)\, q(k)\, e^{s\rho(j,k)}\, \rho(j,k). \tag{7.23}$$

Since $I(Q)$ has been minimised

$$R(D) = I(Q)$$

$$= \sum_j \sum_k p(j)\, q(k/j) \log \frac{q(k/j)}{q(k)}. \tag{7.24}$$

From

$$\frac{q(k/j)}{q(k)} = \lambda(j)\, e^{s\rho(j,k)} \tag{7.25}$$

follows

$$R(D) = s\, D \log e + \sum_j p(j) \log \lambda(j). \tag{7.26}$$

We have now found an expression for the rate of the source as a function of the permissible distortion D. The solution is found in implicit form via the parameter s, an explicit relation is not possible excepting a few simple cases. A value of s yields a value for D and for $R(D)$ and thus gives a point on the $R(D)$ curve.

It can be shown that this parameter s is proportional to the first derivative of $R(D)$ with respect to D and is therefore connected to the slope of the $R(D)$ curve at a certain point. We have

$$s = \frac{d\, R(D)}{dD} \, / \log e. \tag{7.27}$$

The rate distortion function is a continuous, monotone decreasing function for $0 \leq D \leq D_{\max}$. The parameter s is continuous for $0 < D < D_{\max}$ and is not positive.

Example 7.4
This example is an extension of the previous examples in this chapter. We determine the value of the rate distortion function in the case of $D = 0.45$.

Substitution of the probabilities $q(k)$ (compare Example 7.2) and the distortion (see Example 7.1) into equation (7.18) leads to

$$\lambda(1) = \frac{1}{\sum_k q(k)e^{s\rho(j,k)}} = \frac{5}{2 + e^s + 2e^{3s}} \ .$$

The same value for $\lambda(2)$ is obtained as for $\lambda(1)$. Since also $p(1) = p(2)$, we find with equation (7.23)

$$D = \sum_{j=1}^{2} \sum_{k=1}^{3} \lambda(j)p(j)q(k)e^{s\rho(j,k)}\rho(j,k)$$

$$= \lambda(1)p(1)\left\{ \sum_{k=1}^{3} q(k)e^{s\rho(1,k)}\rho(1,k) + \sum_{k=1}^{3} q(k)e^{s\rho(2,k)}\rho(2,k) \right\}$$

$$= \frac{s/2}{2 + e^s + 2e^{3s}} \left\{ 2(\tfrac{1}{5}e^s + \tfrac{6}{5}e^{3s}) \right\}$$

$$= \frac{e^s + 3e^{3s}}{2 + e^s + 2e^{3s}}.$$

The rate distortion function as function of D and s becomes with the help of equation (7.26)

$$R(D) = s D \log e + \sum_j p(j) \log \lambda(j)$$

$$= s D \log e + \log \lambda(1)$$

$$= s D \log e + \log \left\{ \frac{5}{2 + e^s + 2e^{3s}} \right\}.$$

However, we are interested in the value of $R(D)$ for $D = 0.45$. See Figure 7.4.

Figure 7.4. The rate distortion function of Example 7.4.

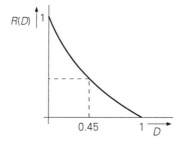

Using $x = e^s$ in the expression for D leads to

$$D = \frac{x + 3x^3}{2 + x + 2x^3} \, ,$$

and thus

$$x + 3x^3 = 0.45(2 + x + 2x^3),$$

or

$$2.1x^3 + 0.55x - 0.9 = 0.$$

The solution of this equation is $x \approx 0.64$. Therefore $s = \ln 0.64 \approx -0.45$. Now the value of $R(0.45)$ can be found.

$$R(D) = s D \log e + \log \left\{ \frac{5}{2 + e^s + 2e^{3s}} \right\}$$

$$= -0.45 \times 0.45 \times \log e + \log \left\{ \frac{5}{2 + 0.64 + 2(0.64)^3} \right\}$$

$$= \frac{1}{\ln 2} (-0.20 + 0.46)$$

$$= \frac{0.26}{\ln 2} \approx 0.37$$

$$\Rightarrow \quad R(0.45) \approx 0.37. \qquad \qquad \triangle$$

Determining the rate distortion function is generally not easy. Because of this a lower limit is often used for the rate distortion function. Numerical methods for determining the $R(D)$ function also exist.

7.3 The binary case

An important application concerns the rate distortion function for a binary source with the previously mentioned distortion measure. In this case, an explicit expression may be found, which can be derived as follows. We assume that the source generates symbols with the probabilities $p(u_1) = p$ and $p(u_2) = 1 - p$. The distortion is as given by the diagram in Figure 7.1.

We further assume without loss of generality that $0 \le p \le \frac{1}{2}$. It is easy to see that $D_{min} = 0$. This occurs if we choose for the transition matrix Q,

$$Q = \begin{bmatrix} 1 & 0 \\ 0 & 1 \end{bmatrix}.$$

The rate distortion function $R(0)$ is equal to $H(U)$ in this case.

$$R(0) = \min_{Q \in Q_D} I(Q)$$

$$= \min_{Q \in Q_D} \{H(U) - H(U/\hat{U})\}$$

$$= H(U) = -p \log p - (1-p) \log (1-p),$$

since the equivocation is zero.

The maximum distortion is $D = p$. This is the best choice that can be made if $I(Q) = 0$, because the alternative $D = 1 - p$ always yields a larger distortion, since $p \leq \frac{1}{2}$. The matrix Q of transition probabilities in this case is

$$\begin{bmatrix} 0 & 1 \\ 0 & 1 \end{bmatrix}.$$

In order to determine the other points of the $R(D)$ curve, the probabilities $q(k)$ must first be found.

From

$$\sum_j \lambda(j) \, p(j) \, e^{s\rho(j,k)} = 1,$$

and assuming $e^s = a$, it follows that

$$\lambda(0){\cdot}p \quad + \lambda(1){\cdot}(1{-}p){\cdot}a = 1,$$

$$\lambda(0){\cdot}p{\cdot}a + \lambda(1){\cdot}(1{-}p) \quad = 1,$$

hence

$$\lambda(0) = \frac{1}{p(1+a)},$$

$$\lambda(1) = \frac{1}{(1-p)\,(1+a)}.$$

Next we determine $q(k)$ from $\lambda(j)$. This can be done on the basis of the relation

$$\lambda(j) = \frac{1}{\sum_k q(k) e^{s\rho(j,k)}}.$$

We find

$$q(0) + a\,q(1) = \frac{1}{\lambda(0)} = p(1+a),$$

$$a\,q(0) + q(1) = \frac{1}{\lambda(1)} = (1-p)\,(1+a)\,,$$

from which follow

$$q(0) = \frac{p - a(1-p)}{1 - a},$$

$$q(1) = \frac{1 - p - ap}{1 - a}\,.$$

If this result is substituted, we find for the permissible distortion

$$D = \frac{a}{1 + a}\,.$$

For $R(D)$ we have

$$R(D) = s\,D \log e + \sum_{j} p(j) \log \lambda(j).$$

Figure 7.5. The rate distortion function for the binary case.

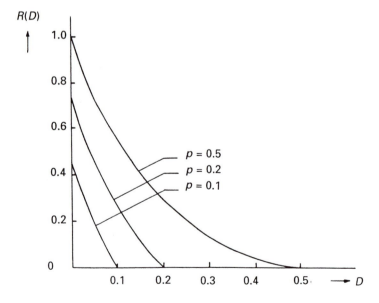

Because

$$a = \frac{D}{1-D},$$

and

$$s = \frac{\log a}{\log e},$$

this can be reduced to

$$R(D) = -p \log p - (1-p) \log (1-p) + D \log D + (1-D) \log (1-D)$$

$$= H(U) - H(D). \tag{7.28}$$

We have thus found a relation, from which we can directly determine the mutual information for a given value of D, which must be conveyed in order to achieve an average distortion D.

Figure 7.5 gives the $R(D)$ curve for a few values of p. From the figure it can be seen that a smaller average distortion can only be achieved by increasing $R(D)$. Also, the rate distortion function $R(D)$ at $p = 0.5$ is larger than for $p < 0.5$ for every value of D, which is intuitively true.

Each point on the curve is reached by a matrix of transitional probabilities which gives rise to both an average distortion $d(Q) = D$ and a mutual information $R(D)$. For the transition probabilities we have

$$q(k/j) = \lambda(j) \, q(k) \, e^{s\rho(j,k)},$$

which yields the following matrix Q:

$$Q = \begin{bmatrix} \dfrac{a(p-1)+p}{p(1-a^2)} & \dfrac{-a(p-1)-a^2p}{p(1-a^2)} \\[3mm] \dfrac{ap-a^2(1-p)}{(1-p)(1-a^2)} & \dfrac{(1-p)-ap}{(1-p)(1-a^2)} \end{bmatrix}. \tag{7.29}$$

7.4 Source coding and information transmission theorems

In this section we will ascertain if a coding theorem can be found for the case where a certain average distortion D is permitted. We shall see that there is a code which, with an average distortion D, transmits information, when the rate distortion function $R(D)$ is smaller than, or at least equal to, the channel capacity C. In a manner of speaking the rate distortion function therefore fulfils the same role as the amount of information. As a consequence the rate distortion function may be regarded as the effective amount of information that the information source generates if the destination permits a certain average distortion D. In the previous section we found that for perfect reproduction $R(D) = R(0) = H(U)$, so that we may regard the amount of information as a special case of the rate distortion function, namely for $D = 0$.

This notion of an effective amount of information is particularly interesting because we are now able to reflect upon the question of what information must be transmitted. Much more is known about the next question, namely how this must be realised. We try to reduce the amount of information generated by a source through the use of various data reduction methods. With speech, for example, attention is paid to intelligibility at the destination, with image signals the recognisability etc. In the end these methods come down to removing the information which is irrelevant for the source and retaining the effective information. A certain amount of distortion is knowingly introduced in this manner, in such a way, however, that the destination receives the relevant part of the amount of information generated by the source (data reduction). In order to come to a formulation of the relationship mentioned between $R(D)$ and C, we must consider the process of data reduction in more detail. Because the original information source may be regarded as replaced by a new source, namely the reproduction, which generates a certain amount of effective information, we will call this process *source coding*.

We will consider source words u and destination words (e.g. code words) $\hat{u}$ chosen from the source and reproducing alphabets respectively, each consisting of L symbols. We call a group B of N different code words a code of size N and length L. For each code word u we choose the destination word $\hat{u} \in B$ for which the distortion $\rho(u,\hat{u})$ between u and $\hat{u}$ is the smallest. Let this destination word be denoted by $\hat{u}_B$. The resulting distortion for a given source word u is then

$$\rho(u,\hat{u}_B) = \min_{\hat{u} \in B} \rho(u,\hat{u}). \tag{7.30}$$

This distortion also depends on the group B, another group may yield another value. The average distortion ρ_B that arises if an arbitrary source word u is presented is now

$$\rho_B = E\left[\rho(u,\hat{u}_B)\right] = \sum_u p(u) \min_{\hat{u} \in B} \rho(u,\hat{u}). \tag{7.31}$$

The *code rate R* is defined as

$$R = \log N \qquad \text{bits/word,}$$

or

$$= \frac{1}{L} \log M \quad \text{bits/symbol.} \tag{7.32}$$

This is, in fact, the maximum amount of information per symbol with respect to a source code of size N and block length L, if the probabilities of all code words are the same.

We now introduce the notion *D-admissibility*. We say that a code B, thus a group of code words, is D-admissible if $\rho_B \leq D$. The smallest size of a D-admissible code is also of importance. We denote this by $N(L,D)$, since apart from the number of symbols L, the size also depends on the permissible average distortion D. We now come to a theorem which is known as the *source coding theorem*.

Theorem 7.1 (Source coding theorem)
For any $\varepsilon > 0$ and any $D \geq 0$ an integer L can be found such that there exists a $(D + \varepsilon)$-admissible code with block length L and with code rate $R < R(D) + \varepsilon$.
In other words, the inequality

$$\frac{1}{L} \log N(L, D + \varepsilon) < R(D) + \varepsilon \tag{7.33}$$

holds for sufficiently large L.   □

The proof of this theorem will not be given here.

We will just prove the converse source coding theorem which states that there is no D-admissible source code with a rate less than $R(D)$.

Theorem 7.2 (Converse source coding theorem)
No D-admissible code has a rate less than $R(D)$. That is, for all n

$$\frac{1}{L} \log N(L,D) \geq R(D). \tag{7.34}$$

Proof
Let $B = (\hat{u}_1, \hat{u}_2, \ldots, \hat{u}_N)$ be a D-admissible code. Let $I(\hat{U};U)$ denote the average mutual information that results when each u is encoded into the $\hat{u} \in B$ for which $\rho(u,\hat{u})$ is minimised, i.e. $\rho(u,\hat{u}_B)$ (compare equation (7.30)). Because encoding is a deterministic process it follows that $H(\hat{U}/U) = 0$ and thus

$$I(\hat{U};U) = H(\hat{U}) \leq \log N. \tag{7.35}$$

It is also the case that

$$I(\hat{U};U) = H(U) - H(U_1,\ldots U_L/\hat{U}_1,\ldots,\hat{U}_L). \tag{7.36}$$

With the help of formulae (3.8) and (3.11) this yields

$$H(U_1,\ldots U_L/\hat{U}_1,\ldots,\hat{U}_L) \leq \sum_{i=1}^{L} H(U_i/\hat{U}_1,\ldots,\hat{U}_L)$$

$$\leq \sum_{i=1}^{L} H(U_i/\hat{U}_i). \tag{7.37}$$

Since the source is memoryless, it is the case that

$$H(U_1,\ldots,U_L) = \sum_{i=1}^{L} H(U_i), \tag{7.38}$$

and thus with equation (7.35) we find

$$\frac{1}{L} \left\{ \sum_{i=1}^{L} H(U_i) - H(U_i/\hat{U}_i) \right\} \leq \frac{1}{L} \log N. \tag{7.39}$$

If D_i is the average distortion with which the ith symbol is reproduced then

$$R(D_i) \leq I(U_i;\hat{U}_i) = H(U_i) - H(U_i/\hat{U}_i). \tag{7.40}$$

Combining equation (7.39) and (7.40) yields

$$\frac{1}{L} \sum_{i=1}^{L} R(D_i) \leq \frac{1}{L} \log N. \tag{7.41}$$

Since B is D-admissible $L^{-1} \sum_i D_i \leq D$ holds. The convexity of the $R(D)$ makes

$$R(D) \leq R \left(\frac{1}{L} \sum_{i=1}^{L} D_i \right) \leq \frac{1}{L} \sum_{i=1}^{L} R(D_i). \tag{7.42}$$

Combination of formulae (7.41) and (7.42) proves the theorem. $\square$

Theorem 7.1 thus says that there is a source code which makes it possible to encode the information source with an average distortion that is a little larger than D, whereby the source code gives a rate which approaches arbitrarily close to the rate distortion function $R(D)$. However, the rate can not become smaller than $R(D)$ without introducing more distortion, as is evident from Theorem 7.2.

On the basis of this theorem, we can say that $R(D)$ may indeed be regarded as the smallest code rate for a permissible average distortion D. That is, it gives an indication of the minimum number of code words.

The source coding theorem only states that a code is possible but does not directly indicate how this can be achieved. We shall now use an example to illustrate various aspects. Consider a binary memoryless symmetric source with symbols 0 and 1. For the distortion measure we take the distortion measure $\rho(u,\hat{u})$ for which $\rho(u,\hat{u}) = 1$ if $u \neq \hat{u}$ and $\rho(u,\hat{u}) = 0$ if $u = \hat{u}$. We now consider a source encoding method where one binary symbol (at a fixed place) is omitted per message of length L and added again at the reproducing side by simply guessing with $p(0) = p(1) = \frac{1}{2}$.

After receipt of the transmitted signal the uncertainty or amount of information is just 1 bit: the removed symbol was either 0 or 1. That means $H(U/\hat{U}) = 1$ bit. The maximum amount of transported information is then

$$R^*(D) = H(U) - H(U/\hat{U})$$

$$= \log 2^L - 1 = L - 1 \text{ bits/message}$$

$$= \frac{L-1}{L} \text{ bits/symbol.} \tag{7.43}$$

For the distortion we find

$$D = \frac{1}{L} \left(\tfrac{1}{2} \cdot 0 + \tfrac{1}{2} \cdot 1\right) = \frac{1}{2L} \ ,$$

and thus

$$R^*(D) = 1 - 2D. \tag{7.44}$$

The rate distortion function for a binary source with equiprobable symbols is equal (compare equation (7.28)) to

$$R(D) = H(U) - H(D) = 1 - H(D) \tag{7.45}$$

The results are shown in Figure 7.6. It can easily be seen that reproduction without distortion is only possible for $L \to \infty$, since $R^*(D) \to 1$ and $D \to 0$. In practice, more subtle methods are usually employed.

Let us reconsider Theorem 7.1. The theorem states that a $(D+\varepsilon)$-admissible code exists that guarantees a mapping from source words u to code words $\hat{u}$ with distortion less than or equal to $(D+\varepsilon)$. At the same time the rate of the encoder output is at most

$$R = \frac{1}{L} \log N(L,D+\varepsilon).$$

On the basis of the same theorem we know that R may approach $R(D)$ arbitrarily closely. Since Theorem 7.1 guarantees that we can make $R < R(D) + \varepsilon$, Shannon's second channel coding theorem, Theorem 4.5, implies that the source encoder output can be recovered at the decoder on the other side of the channel with an arbitrary error probability provided that the channel has a capacity $C > R(D) + \varepsilon$. This leads to the following theorem.

Theorem 7.3 (Information transmission theorem)
For all $\varepsilon > 0$ the output of an information source can be reproduced with a distortion of at most D at the output of any discrete memoryless channel of capacity C, provided that

$$C > R(D) + \varepsilon. \tag{7.46}$$

$\square$

One can thus venture that a communication system is ideal in the information theoretic sense if it achieves an average distortion D for this value of $R(D)$ where $R(D) = C$. This conclusion has an important consequence, namely we can now determine what the average distortion is which can be achieved with a given source and a given channel by equalising the rate distortion function of the source with the channel capacity. Because it is desired that $R(D) \leq C$ and because $R(D)$ decreases with increasing D, this is also the smallest average distortion that is possible.

Figure 7.6. Comparison of a source coding method with the rate distortion function.

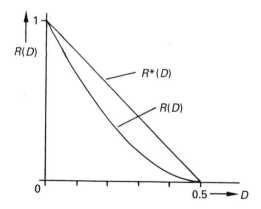

Hence we have found a lower bound for the average distortion, which is known as the *rate distortion bound*.

Here a direct relation is made between the properties of the encoder, the capacity of the channel, and the introduced distortion of the receiving side. There is also a converse information transmission theorem which states that it is impossible to reproduce the source information with a maximal permitted distortion D at the receiving end of any channel with capacity $C < R(D)$.

Thus the practical importance is that if a system designer is required to have a distortion of at most D and the channel has a capacity C he can only possibly succeed if $C \geq R(D)$.

7.5 The continuous rate distortion function

The most important application of the rate distortion theory lies with the reproduction of a continuous information source. The reason for this is that it is quite possible to obtain a distortion-free reproduction with a discrete source, while this is by definition impossible for a continuous source and usually also unnecessary. With the quantisation of continuous signals especially, the situation arises where a certain amount of quantisation distortion is accepted in order to keep down the necessary number of bits.

We shall now examine the definition of the rate distortion function for the continuous case in more detail. There is a clear similarity with the discrete case.

Consider a continuous memoryless source with probability density $p(x)$. We call the conditional probability density of $\hat{x}$ given the value x $q(\hat{x}/x)$ and the probability density at the recovering side $q(\hat{x})$. For the distortion measure we take $\rho(x,\hat{x})$, which gives the distortion when reproducing the value $\hat{x}$ from the value x. The choice of this measure will be explained further. We may now define the average distortion for a given conditional probability density $q(\hat{x}/x)$ as

$$d(q) = \int_{-\infty}^{\infty} \int_{-\infty}^{\infty} p(x) \, q(\hat{x}/x) \, \rho(x,\hat{x}) \, dx \, d\hat{x}. \tag{7.47}$$

For the average mutual information we also have

$$I(q) = \int_{-\infty}^{\infty} \int_{-\infty}^{\infty} p(x) \, q(\hat{x}/x) \log \frac{q(\hat{x}/x)}{q(\hat{x})} \, dx \, d\hat{x}. \tag{7.48}$$

The rate distortion function $R(D)$ is the minimum of $I(q)$ for which the distortion $d(q)$ is at most equal to the permissible distortion D. For a continuous system this leads to the definition

$$R(D) = \inf_{q \in q_D} I(q), \tag{7.49}$$

where q_D is the set of conditional probability density functions which satisfy $d(q) \leq D$.

The calculation of the rate distortion function proceeds analogously to that of the discrete case. We find

$$q(\hat{x}/x) = \lambda(x) \, q(\hat{x}) \, e^{s\rho(x,\hat{x})}, \tag{7.50}$$

with

$$\lambda(x) = \dfrac{1}{\displaystyle\int_{-\infty}^{\infty} q(\hat{x}) \, e^{s\rho(x,\hat{x})} \, d\hat{x}}, \tag{7.51}$$

from which follows

$$\int_{-\infty}^{\infty} \lambda(x) \, p(x) \, e^{s\rho(x,\hat{x})} \, dx = 1. \tag{7.52}$$

This results finally in

$$D = \int_{-\infty}^{\infty} \int_{-\infty}^{\infty} \lambda(x) \, p(x) \, q(\hat{x}) \, e^{s\rho(x,\hat{x})} \, \rho(x,\hat{x}) \, dx \, d\hat{x}, \tag{7.53}$$

and

$$R(D) = s \, D \log e + \int_{-\infty}^{\infty} p(x) \log \lambda(x) \, dx. \tag{7.54}$$

The relationship between $R(D)$ and D is thus obtained in parametric form here as well. The parameter s is proportional to the derivative of $R(D)$, thus with the slope of the $R(D)$ curve.

A number of the derived properties for the discrete source remain valid. We may state that

$$D_{\min} = 0, \tag{7.55}$$

$$D_{\max} = \inf_{\hat{x}} \int_{-\infty}^{\infty} p(x) \, \rho(x,\hat{x}) \, dx. \tag{7.56}$$

Further, $R(D)$ is a monotonically decreasing non-negative function for $0 < D < D_{max}$. A difference with the discrete source is that the $R(D)$ curve goes to infinity for $D \to 0$. A smaller distortion means that we wish to reproduce the source in more detail. More bits are then necessary in order to specify a value, and these must be transported. This means that for $D \to 0$, $R(D)$ goes to infinity. This is of course a physical impossibility.

The distortion that is introduced with continuous sources is a function of the difference between the values generated at the source and reconstructed at the destination. An often used distortion measure is the squared-error criterion, which reads

$$\rho(x,\hat{x}) = (x - \hat{x})^2. \tag{7.57}$$

A consequence of this definition is that large errors weigh more heavily. The absolute error is also used as distortion measure, that is

$$\rho(x,\hat{x}) = |x - \hat{x}|. \tag{7.58}$$

Without giving a derivation, we note that for an information source with a gaussian probability density $N(0,\sigma^2)$, the rate distortion function – if we choose the squared error as distortion measure – is given by

$$\left.\begin{array}{ll} R(D) = \dfrac{1}{2} \log \dfrac{\sigma^2}{D} \text{ bits/sample} & \text{for } 0 \le D \le \sigma^2, \\[2mm] \quad\quad = 0 & \text{for } D > \sigma^2, \end{array}\right\} \tag{7.59}$$

where σ^2 is the power of the information source per sample. The $R(D)$-curve is sketched in Figure 7.7.

The maximum allowable distortion D_{max} may be determined as follows. If no information is transmitted, the best choice which the receiver can make consists of choosing the average value (here equal to zero) because this has the largest probability density and is therefore the most probable transmitted value. The error which is now made is $(x - \sigma)^2 = x^2$ and the average error $d(q)$ is $E\{(x - \sigma)^2\} = E\{x^2\} = \sigma^2$. From this it then follows that D_{max}, which is the minimum distortion for $I(q) = R(D) = 0$, is equal to σ^2.

We further see that $R(D)$ increases to infinity if D is made smaller and smaller. This is of course a mathematical abstraction; in practical systems D will be bounded by the smallest measurable signal value.

If the information source is constrained in the bandwidth W, the source may be determined by $2W$ independent samples per second. Hence

$$R(D) = W \log \frac{\sigma^2}{D} \text{ bits/sec} \quad \text{for } 0 \le D \le \sigma^2,$$
$$= 0 \qquad\qquad\quad \text{for } D > \sigma^2. \tag{7.60}$$

In the foregoing we introduced the concept of the rate distortion bound, it being the average distortion for which $R(D) = C$. It is the smallest average distortion that is possible.

This rate distortion bound will be derived for the case where we connect a gaussian information source with bandwidth W_x and power σ_x^2 per sample to a channel that has a bandwidth W_c and is perturbed by additive white gaussian noise with power σ_n^2 per sample. For the source we have

$$R(D) = W_x \log \frac{\sigma_x^2}{D} \text{ bits/sec} \quad \text{for } 0 \le D \le \sigma_x^2. \tag{7.61}$$

The channel has as capacity

Figure 7.7. The rate distortion function for the gaussian case.

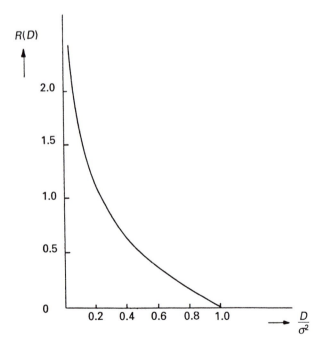

$$C = W_c \log (1 + \frac{P_x}{P_n}),$$

where P_x is the power of the input signal in the channel. We find the rate distortion bound by equalising $R(D)$ and C. This gives

$$D = \sigma_x^2 \left\{ 1 + \frac{P_x}{2 \, W_c \, \sigma_x^2} \right\}^{-\frac{W_c}{W_x}} \qquad \text{for } 0 \leq D \leq \sigma_x^2. \qquad (7.62)$$

This is thus the smallest possible distortion which can be achieved. If we consider the case of ideal adjustment, $W_c = W_x = W$ and $P_x = 2W \, \sigma_x^2$, then it follows that

$$D' = \frac{\sigma_x^2}{1 + \frac{\sigma_x^2}{\sigma_n^2}}. \qquad (7.63)$$

If $\sigma_n^2 \gg \sigma_x^2$, then $D' \approx \sigma_x^2$, which means that the channel is so bad that we can in practice make do with continually choosing the mean value of the source signal. If $\sigma_x^2 \gg \sigma_n^2$, then $D' \approx \sigma_n^2$. This too is clear, the receiver cannot perceive the signal of the information source more accurately than the noise in the channel allows because σ_n^2 is the smallest perceptible power per sample. If the source is connected directly to the channel, the average distortion is equal to this smallest perceptible power σ_n^2.

For the derivation of the rate distortion function it was always assumed that $0 \leq D \leq D_{\max}$. In this example it has been shown that for $\sigma_n^2 \gg \sigma_x^2$ the average distortion approaches $D_{\max}$, so that without a source we could in actual fact expect the same result at the receiving end. D did not go to zero for $\sigma_x^2 \gg \sigma_n^2$, but remained σ_n^2. To be able to achieve an average distortion $D < \sigma_n^2$ it seems one could strive towards a system where the bandwidth of the channel is larger than the bandwidth of the information source, thus $W_c > W_x$, which means that one must pursue a certain bandwidth-expansion. For $W_c \gg W_x$ the average distortion D will go to zero.

This example illustrates an application of the rate distortion bound, where one can find an expression for the minimum average distortion that can be achieved with a given source and a given channel. Moreover, using the expression found, it emerges that indications can be found as to how one can improve a communication system.

7.6 Exercises

7.1. Given a binary source which generates symbols u_0 and u_1 from the alphabet U. The source symbols occur with the probabilities $p(u_0) = p$ and $p(u_1) = 1 - p$ $(0 \le p \le \frac{1}{2})$. The Z-channel is given as well (Figure 7.8). $\{v_0, v_1\}$ are the destination symbols from the alphabet V. The transition probabilities $q(v_j/u_i)$ are given by

$$q(v_0/u_0) = 1, \quad q(v_1/u_1) = 1 - q.$$

The distortion between source and reproducing symbols $\rho(u_i, v_j)$ is given by

$$\rho(u_0, v_0) = \rho(u_1, v_1) = 0,$$

$$\rho(u_1, v_0) = \rho(u_0, v_1) = 1.$$

a. Calculate the average distortion $d(Q)$.
b. Give the definition of the rate distortion function $R(D)$. What is the maximal value of $R(D)$? For what value of q is this value obtained? How large is the average distortion $d(Q)$ in this case?
c. What is the minimal value of $R(D)$? For what value of q is this value obtained? How large is the average distortion $d(Q)$ in this case?
d. Sketch $R(D)$ as a function of D.

7.2. Consider a channel with input alphabet (x_1, x_2), output alphabet (y_1, y_2) and transition probabilities $q(y_j/x_i)$. At the input the probability of x_1 equals $\frac{1}{4}$. For the symbol distortion

$$\rho(x_1, y_1) = 0, \quad \rho(x_1, y_1) = \alpha, \quad \rho(x_2, y_1) = 5 - \alpha, \quad \rho(x_2, y_2) = 0,$$

with $0 \le \alpha \le 5$.
In a and b it is assumed that $q(y_1/x_1) = \frac{3}{5}$ and $q(y_1/x_2) = \frac{3}{10}$.
a. Calculate the amount of information at the output of the channel.
b. Calculate the average distortion as a function of α. What is the smallest average distortion obtainable?
Now consider the rate distortion function $r(D)$ of the given system.

Figure 7.8. The Z-channel of Exercise 7.1.

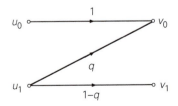

c. Calculate $R(0)$ and give the corresponding channel matrix.

d. Calculate D_{max}, the smallest distortion possible when no information from the source is received. What value of α will obtain the largest D_{max}?

7.7 Solutions

7.1. *a.* For the joint probabilities $p(u_i, v_j)$ we find

$$p(u_0, v_0) = p(v_0/u_0)\, p(u_0) = p,$$
$$p(u_0, v_1) = p(v_1/u_0)\, p(u_0) = 0,$$
$$p(u_1, v_0) = p(v_0/u_1)\, p(u_1) = q(1 - p),$$
$$p(u_1, v_1) = p(v_1/u_1)\, p(u_1) = (1 - q)(1 - p).$$

Hence the average distortion becomes

$$d(Q) = \sum_{i=0}^{1} \sum_{j=0}^{1} p(u_i, v_j)\rho(u_i, v_j)$$

$$= q(1 - p).$$

b. For the rate distortion function it is the case that

$$R(D) = \min_{Q \in Q_D} I(Q) \text{ bits/symbol},$$

where $I(Q)$ is the mutual information and

$$Q_D = \{Q \mid d(Q) \le D).$$

The rate distortion function is maximally equal to

$$\max R(D) = H(U).$$

This maximum occurs if there is a one-to-one relation between source and reproducing symbols. Concerning the transition probabilities this implies

$$q(v_0/u_0) = q(v_1/u_1) = 1$$

and thus $1 - q = 1$ and $q = 0$. Now, the average distortion becomes $d(Q) = 0$.

c. The minimal value of the mutual information $I(Q)$ is equal to 0 and thus the minimal value of the rate distortion function also equals 0. This minimum appears if the reproducing symbols give no information about the original source symbols. This is the case if both source symbols u_0

and u_1 give rise to the same reproducing symbol. This leads to $q = 1$ and for the corresponding average distortion we obtain $d(Q) = 1 - p$.

d. See Figure 7.9.

7.2. See Figure 7.10.

Note. The pair-wise distortions are given within brackets.

a. For the marginal probabilities $q(y_1)$ and $q(y_2)$ we find

$$q(y_1) = \sum_{i=1}^{2} p(x_i)q(y_j/x_i) = \frac{1}{4}\frac{3}{5} + \frac{3}{4}\frac{3}{10} = \frac{3}{8},$$

$$q(y_2) = 1 - \frac{3}{8} = \frac{5}{8}.$$

Now, the amount of information $H(Y)$ becomes

$$H(Y) = -\frac{3}{8}\,\log\frac{3}{8} - \frac{5}{8}\,\log\frac{5}{8} = 0.954 \text{ bit.}$$

b. The average distortion can be found easily:

$$d(Q) = \sum_{i=1}^{2} p(x_i)p(y_j/x_i)\rho(x_i,y_j)$$

$$= \frac{1}{4}\frac{2}{5}\cdot\alpha + \frac{3}{4}\frac{3}{10}\,(5 - \alpha) = -\frac{1}{8}\alpha + \frac{9}{8}.$$

Figure 7.9. The rate distortion function of Exercise 7.1.

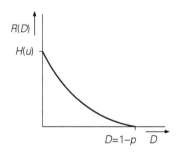

Figure 7.10. Distortions and transition probabilities of Exercise 7.2.

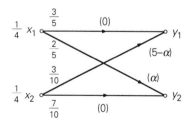

The average distortion has a minimum for $\alpha = 5$:

$$\min_{\alpha} d(Q) = \tfrac{1}{2}.$$

c. If $D = 0$, the rate distortion function achieves a maximum equal to $H(X)$, which is the information of the source,

$$H(X) = -\tfrac{1}{4} \log \tfrac{1}{4} - \tfrac{3}{4} \log \tfrac{3}{4} = 0.811.$$

The corresponding channel matrix is

$$Q = \begin{bmatrix} 1 & 0 \\ 0 & 1 \end{bmatrix}.$$

That means, there is a one-to-one correspondence between input and output symbols.

d. On the basis of the theory it is the case that

$$D_{\max} = \min_{j} \sum_{i=1}^{2} p(x_i)\rho(x_i, y_j).$$

In the present case, we obtain

$$D_{\max} = \min \left(\tfrac{1}{4}\, \alpha, \tfrac{3}{4}\, (5 - \alpha) \right)$$

and thus

$$D_{\max} = \tfrac{1}{4}\, \alpha \qquad \text{for } 0 \le \alpha \le \tfrac{15}{4},$$

$$= \tfrac{3}{4}\, (5 - \alpha) \qquad \text{for } \tfrac{15}{4} \le \alpha \le 5.$$

For $\alpha = \tfrac{15}{4}$, $D_{\max}$ achieves its absolute maximum $\left(= \tfrac{15}{16} \approx 0.938 \right)$.

Figure 7.11. The rate distortion function of Exercise 7.2.

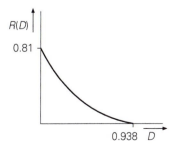

8

Network information theory

8.1 Introduction

Up till now, our discussion of point-to-point communication was only concerned with the case in which the transmitter and receiver are connected by a private communication channel. In practice, however, the necessity of cost-reduction will force the transmitter and receiver to use a communication channel which can be accessed by a large number of users.

In general, *multiterminal communication networks* consist of more than one transmitter or receiver, or even more than one channel. An example is given by a *multi-access communication network*, which has several transmitters and receivers, which all access the same channel. If there are several transmitters, but only one receiver, the term *multi-access communication channel* is used. For instance, a number of satellite ground stations, which all send their information to the same satellite. One of the most important aspects of a multi-access communication network is to ensure that the channel is always used optimally. On the one hand, it is desirable to exploit the capacity of the channel to the fullest, whereas on the other hand, the access time for the users must remain as short as possible. The most straightforward solution to this is circuit switching. With the exception of a single transmitter and receiver, the channel is closed for all other users. Thus, the problem has in fact been reduced to a simple point-to-point connection. However, it is clear that this method is not particularly efficient. A pause during communication will immediately result in inefficient usage of the channel. A better method can be found by multiplexing, where the information of several sources is combined into a single data stream.

If there are a number of receivers, but only one transmitter, then the phrase *broadcast channel* is normally used. A well-known example is given by a TV transmitter, which sends programmes to TV receivers all over the world.

A specific type of communication channel is the *two-way channel*. This has two terminals, which can adopt the role of both transmitter and receiver, simultaneously. This means that the information stream in one direction will coincide with the information stream in the other direction.

The problem of calculating the capacity of networks in general terms still remains unsolved. This also applies to finding the most efficient method of encoding. A satisfactory solution to these problems can only be found in certain very specific cases. An adequately comprehensive theory has still not been found.

In this chapter we will examine the information theoretical aspects of several special networks, such as the multi-access communication channel, the broadcast channel and the two-way channel.

8.2 Multi-access communication channel

A multi-access communication channel comprises a number of transmitters and exactly one receiver. Consider Figure 8.1. All of the transmitters use the same channel. The information reaching the receiver is equal to the sum of the, say k, signals of the transmitters, plus the noise introduced by the channel. One method of tackling this problem is to divide the channel into a number of sub-channels and to assign a sub-channel to each of the transmitters. Now we will in fact have a number of point-to-point communication links. The division into sub-channels can be accomplished in various ways, for instance by *frequency-division multi-access* and *time-division multi-access*.

Figure 8.1. Multi-access communication channel.

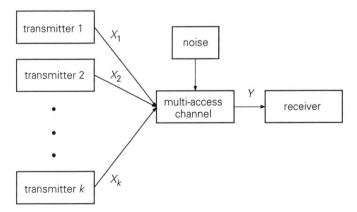

In the case of multi-access by means of frequency-division, each user is assigned a well-defined section of the available frequency band for sending information to the receiver. In equation (6.15) we found an expression for the capacity of a channel inflicted with additive white gaussian noise with an average power P_n. If P_x denotes the average input power to the channel, then

$$C = W \log \left\{ 1 + \frac{P_x}{P_n} \right\}. \tag{8.1}$$

If the total frequency band W is divided into k equal sections, then each transmitter may use a bandwidth W/k, instead of W. Substituting this value W/k in equation (8.1) and remembering that the average noise power for each transmitter decreases by a factor k, we find the following expression for the channel capacity that each user sees:

$$C' = \frac{W}{k} \log \left\{ 1 + \frac{kP_x}{P_n} \right\}. \tag{8.2}$$

For multi-access by means of time-division, each user may access the channel for a predefined period in time. Therefore, with k users, each user may actually use the channel for $1/k$ of the time. During this time, each user sees a channel with an average power kP_x and therefore a channel with a capacity C equal to that given in expression (8.2).

It is also possible to assign the channel capacity to each of the k users according to the information rates of the other users. If several users exhibit a decreasing rate, then the rate of the other users may be increased, temporarily. Naturally, the rate R_i of a single user i can never exceed the maximum value, if he were to have exclusive access to the channel. In other words,

$$R_i \leq W \log \left\{ 1 + \frac{P_{x_i}}{P_n} \right\}. \tag{8.3}$$

If there are k users, then the combined rate can never be greater than that of the sum of the information streams with a total power

$$\sum_{i=1}^{k} P_{x_i}.$$

Thus, it follows that

$$\sum_{i=1}^{k} R_i \leq W \log \left\{ 1 + \frac{\sum_{i=1}^{k} P_{x_i}}{P_n} \right\}. \tag{8.4}$$

If the users do not collaborate, then from the point of view of a single transmitter, the other transmitters can be regarded as noise sources. If this noise can be considered as gaussian noise, the total noise power will be equal to

$$P_n + \sum_{j=1, j \neq i}^{k} P_{x_j},$$

resulting in a channel capacity

$$C_i = W \log \left\{ 1 + \frac{P_{x_i}}{P_n + \sum_{j=1, j \neq i}^{k} P_{x_j}} \right\}. \tag{8.5}$$

This expression is equal to the achievable rate for this case.

If each transmitter transmits at the same power level, i.e. $\forall j \ P_{x_j} = P_x$, we find

$$C = W \log \left\{ 1 + \frac{P_x}{P_n + (k-1)P_x} \right\}. \tag{8.6}$$

Since it is the case that

$$\ln a \leq a - 1 \text{ and therefore } \ln(a + 1) \leq a \text{ and } \log(a + 1) \leq a \log e,$$

the rate of each user is restricted to

$$R \leq C = W \log \left\{ 1 + \frac{P_x}{P_n + (k-1)P_x} \right\} \leq W \frac{P_x}{P_n + (k-1)P_x} \log e. \tag{8.7}$$

As can be expected, the rate of each user will decrease as the noise level increases, or as the total number of users increases.

If the total number of users is large (k is large), then

$$R \leq \frac{W}{k} \log e. \tag{8.8}$$

Therefore, the rate of each of the k users must satisfy this expression.

In the previous discussion we considered the gaussian case. We will now proceed with a more general case, in which the assumption of gaussian sources is not made, and attempt to find an expression for the capacity of a multi-access channel, or, more precisely, the capacity regions.

We will begin with a theorem which is a generalised version of the data processing theorem (Theorem 4.6). This theorem will play an important role in deriving the necessary conditions for transferring information through a multi-access channel with a sufficiently small error probability.

Theorem 8.1

Assume a cascade of processing steps, in which the input of each processing step is equal to the output of the previous processing step (compare Figure 8.2). Then the mutual information between input and output will decrease for an increasing number of processing steps:

$$I(U;V) \leq I(X;Y). \tag{8.9}$$

Proof

Let U, X, Y and Z represent the inputs/outputs of the various processing steps. Without loss of generality, we will consider the inputs/outputs at the per symbol level.

Let u_l, x_i, y_j and z_k denote symbols of the corresponding alphabets. The term (X,U) represents the joint event, which is characterised by the joint probabilities $p(x_i,u_l)$. Thus, the mutual information between Y and (X,U) will equal

$$I(Y;(X,U)) = \sum_{i,j,l} p(u_l,x_i,y_j) \log \left\{ \frac{p(y_j/x_i,u_l)}{p(y_j)} \right\}$$

$$= \sum_{i,j,l} p(u_l,x_i,y_j) \log \left\{ \frac{p(y_j/x_i,u_l)}{p(y_j/x_i)} \right\} + \sum_{i,j,l} p(u_l,x_i,y_j) \log \left\{ \frac{p(y_j/x_i)}{p(y_j)} \right\}. \tag{8.10}$$

The first term of the lower line of equation (8.10) can be interpreted as the mutual information between U and Y, given X. The second term is in fact the mutual information between X and Y. Therefore, we obtain

$$I(Y;(X,U)) = I(U;Y/X) + I(X;Y).$$

Figure 8.2. Cascading of processing steps.

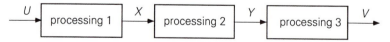

By interchanging X and U, it also follows that

$$I(Y;(X,U)) = I(U;Y/X) + I(X;Y)$$
$$= I(X;Y/U) + I(U;Y). \tag{8.11}$$

Since the output Y only depends on X and not on U, it is the case that

$$\forall i, j, l \;\; p(y_j/x_i, u_l) = p(y_j/x_i).$$

Substituting this expression into $I(U;Y/X)$ leads to

$$I(U;(Y/X)) = 0. \tag{8.12}$$

Therefore

$$I(Y;(X,U)) = I(X;Y) = I(X;Y/U) + I(U;Y) \tag{8.13}$$

and

$$I(X;Y) \geq I(U;Y). \tag{8.14}$$

In the same way it can be shown that

$$I(U;Y) \geq I(U;V). \tag{8.15}$$

Finally, combining these two inequalities results in

$$I(X;Y) \geq I(U;Y) \geq I(U;V), \tag{8.16}$$

with which the correctness of the theorem is demonstrated.   □

The theorem states that processing the information will cause the mutual information between input and output to decrease. In other words, processing will cause the input and output to become more independent.

Let us now consider the situation of Figure 8.3.

The input of encoder 1 is a series of blocks U^N of length N, with symbols selected from an alphabet of size J. The output of the encoder is a series of blocks X^n of length n. In the same manner, for encoder 2 the input and the output blocks are denoted by U'^N and X'^n, respectively. The output of the

Figure 8.3. Multi-access channel with two transmitters.

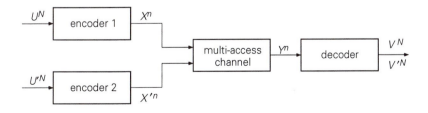

channel, which is considered to be memoryless, is a series of code words Y^n. Finally, the results of decoding two output blocks are represented by V^N and V'^N, which are related to the original blocks U^N and U'^N, respectively.

Considering a multi-access channel with two transmitters, the *capacity region* is defined as the convex hull of all rate pairs (R_1,R_2), with R_1 and R_2 representing the individual rates of user 1 and user 2, respectively, for which codes exist which allow the information to be transferred through the channel with an arbitrarily small error probability. Obviously, if one of the rates is equal to zero, then the other rate is only limited by the maximum capacity of the channel. Therefore, the co-ordinates $(C_1,0)$ and $(0,C_2)$ are located within the capacity region. Furthermore, if $(C_1,0)$ and $(0,C_2)$ lie in the capacity region, then the point $(\lambda C_1,(1 - \lambda)C_2)$ must also be located within the region, for all values $\lambda \in [0,1]$. This becomes evident by realising that if time sharing is applied, user 1 will have access to the channel for a fraction λ of the available time and consequently, user 2 may access the channel for the remaining fraction $(1 - \lambda)$. Naturally, the points below the line connecting the points $(C_1,0)$ and $(0,C_2)$ also represents rates for which an arbitrarily small error probability can be achieved.

The following theorem enables us to clearly define the boundaries of the capacity region.

Theorem 8.2
Assuming the inputs X and X' of the multi-access channel are mutually independent, then if it is the case that, for every joint distribution,

$$R_1 > I(X; Y/X') \tag{8.17}$$

or

$$R_2 > I(X'; Y/X) \tag{8.18}$$

or

$$R_1 + R_2 > I((X; X');Y), \tag{8.19}$$

then no code with a rate (R_1,R_2) exists which can guarantee an arbitrary small error probability after the receiver has decoded the information.

Proof
The proof is based on Fano's inequality.
Considering the information at the symbol level, the relation between the conditional information with respect to the lth output symbol v_l of the

decoder given the lth input symbol u_l of encoder 1 on the one hand, and the error probability on the other, is given by

$$P_{e_1} \log (J-1) + H(P_{e_1}) \geq H(v_l/u_l), \tag{8.20}$$

in which the subscript denotes encoder 1. This follows directly from Fano's inequality (see Theorem 4.4) and from the fact that simply reversing the roles of X and Y in Theorem 4.4 leads to an identical bound for $H(Y/X)$.

By summing both sides over all symbols of the block and taking into account the fact that the information measures and the error probability are concave functions, this leads to

$$P_{e_1} \log (J-1) + H(P_{e_1}) \geq \frac{1}{N} \sum_{l=1}^{N} H(v_l/u_l). \tag{8.21}$$

In the same manner, with respect to encoder 2, we obtain

$$P_{e_2} \log (J'-1) + H(P_{e_2}) \geq \frac{1}{N'} \sum_{l=1}^{N'} H(v_l'/u_l'). \tag{8.22}$$

Clearly, for blocks U^N and V^N of length N it is the case that

$$H(V^N/U^N) \leq \sum_{l=1}^{N} H(v_l/u_l). \tag{8.23}$$

On the other hand though, by using the definition of the mutual information, we can write

$$H(V^N/U^N) = H(V^N) - I(U^N;V^N).$$

Using Theorem 8.1, we now obtain

$$H(V^N/U^N) \ = H(V^N) - I(U^N;V^N)$$

$$\geq H(X^n) - I(X^n;Y^n)$$

$$\geq H(X^n) - I(X^n;(Y^n/X'^n)). \tag{8.24}$$

In the latter inequality, the expression $I(X^n;(Y^n/X'^n))$ represents the mutual information between the output of encoder 1 and the output of the decoder, assuming that the output of encoder 2 is known. The inequality is based on the fact that

$$I(X^n;Y^n) \leq I(X^n;(Y^n/X'^n)). \tag{8.25}$$

This can be explained as follows.

Let x^n, x'^n and y^n denote the different blocks that can occur. It will then follow that

$$I(X^n;Y^n/X'^n) = \sum_{x^n,x'^n,y^n} p(x^n,x'^n,y^n) \log \left\{ \frac{p(y^n/x'^n,x^n)}{p(y^n/x'^n)} \right\}$$

$$= \sum_{x^n,x'^n,y^n} p(x^n,x'^n,y^n) \log \left\{ \frac{p(y^n/x^n)}{p(y^n)} \right\}$$

$$+ \sum_{x^n,x'^n,y^n} p(x^n,x'^n,y^n) \log \left\{ \frac{p(y^n)p(y^n/x'^n,x^n)}{p(y^n/x^n)p(y^n/x'^n)} \right\}$$

$$= I(X^n;Y^n) + \sum_{x^n,x'^n,y^n} p(x^n,x'^n,y^n) \log \left\{ \frac{p(y^n)p(x^n/x'^n,y^n)p(x^n,y^n)/p(x'^n,x^n)}{p(y^n/x^n)p(x'^n,y^n)/p(x^n)} \right\}$$

$$= I(X^n;Y^n) + \sum_{x^n,x'^n,y^n} p(x^n,x'^n,y^n) \log \left\{ \frac{p(y^n)p(x^n/x'^n,y^n)/p(x'^n,x^n)}{p(y^n,x^n)/p(x'^n)p(x^n)} \right\}$$

$$= I(X^n;Y^n) + \sum_{x^n,x'^n,y^n} p(x^n,x'^n,y^n) \log \left\{ \frac{p(x^n/x'^n,y^n)}{p(x^n/y^n)} \right\}$$

$$= I(X^n;Y^n) + I(X'^n;X^n/Y^n).$$

Therefore,

$$I(X^n;Y^n/X'^n) \geq I(X^n;Y^n).$$

Combining equations (8.23) and (8.24) and noting that the rate of encoder 1 is given by $R_1 = H(X^n)/n$, we obtain

$$H(V^N/U^N) \geq H(X^n) - I(X^n;Y^n/x'^n)$$

$$= \sum_{i=1}^{N} H(V_i,U_i) \geq nR_1 - I(X^n;Y^n/X'^n)$$

$$\geq nR_1 - \sum_{l=1}^{n} I(X_l;Y_l/X_l')$$

$$= \sum_{l=1}^{n} [R_1 - I(X_l;Y_l/X_l')]. \tag{8.26}$$

By repeating the same steps for encoder 2, we will arrive at

$$\sum_{i=1}^{N} H(V_i',U_i') \geq \sum_{l=1}^{n} [R_2 - I(X_l';Y_l/X_l)]. \tag{8.27}$$

The sum of these two expressions is given by

$$\sum_{i=1}^{N} H(V_i,U_i) + \sum_{i=1}^{N} H(V_i',U_i') \geq \sum_{l=1}^{n} [R_1 + R_2 - I(X_l,X_l');Y_l)]. \tag{8.28}$$

Thus, with the aid of Fano's inequality, we now have three inequalities:

$$N\{P_{e_1} \log (J-1) + H(P_{e_1})\} \geq \sum_{l=1}^{n} [R_1 - I(X_l;Y_l/X_l')], \tag{8.29}$$

$$N'\{P_{e_2} \log (J'-1) + H(P_{e_2})\} \geq \sum_{l=1}^{n} [R_2 - I(X_l;Y_l/X_l)], \tag{8.30}$$

$$N\{P_{e_1} \log (J-1) + H(P_{e_1})\} + N'\{P_{e_2} \log (J'-1) + H(P_{e_2})\}$$
$$\geq \sum_{l=1}^{n} [R_1 + R_2 - I((X_l,X_l');Y_l)]. \tag{8.31}$$

If one of the expressions to the right of the larger than or equal to signs remains positive for every value of *l*, then the error probability cannot become zero. In other words, it will not be possible to decode the information without errors. This statement is identical to what the theorem implies. □

The converse theorem can be understood without too much difficulty and therefore we will omit a complete proof.

Theorem 8.3
Assuming the block length *n* is sufficiently large, then suitable codes can be found for the multi-access channel with two transmitters with rates R_1 and R_2 and a sufficiently small error probability, provided

$$R_1 \leq I(X;Y/X'), \tag{8.32}$$

$$R_2 \leq I(X';Y/X) \tag{8.33}$$

and $$R_1 + R_2 \leq I(X;X');Y) \tag{8.34}$$

for some joint distribution of the inputs X and X', with X and X' mutually independent. ☐

Considering Theorems 8.2 and 8.3, we can conclude that the capacity region of the memoryless channel covers the area enclosed by the convex limit of all the rates R_1 and R_2 which satisfy the three conditions of Theorem 8.3, for a certain product probability distribution of the input pair (X,X').

For a specific product probability distribution, the capacity region will resemble Figure 8.4.

Point a corresponds to the situation in which transmitter 1 is not sending any information to the receiver and transmitter 2 is sending information with a maximum rate. This implies that

$$\max R_2 = \max_{X,X'} I(X';Y/X).$$

For every product distribution within (X,X'), it is the case that

$$I(X';Y/X) = \sum_x p(x) I(X';Y/X=x)$$

$$\leq \max_x I(X';Y/X=x).$$

This maximum can be found by searching for the distribution X' for which the mutual information reaches its maximum.

Therefore,

$$\max R_2 \leq \max_{X'} \max_x I(X';Y/X=x). \tag{8.35}$$

Point b corresponds to the maximum rate at which transmitter 1 can send its information to the receiver, if transmitter 2 is already sending at the maximum rate possible. Transmitter 1 can regard the information sent by transmitter 2 to the receiver as noise. Our previous discussion demonstrated

Figure 8.4. Achievable capacity region of a multi-access channel.

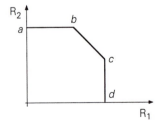

that the maximum rate transmitter 1 can achieve in this case is equal to $I(X;Y)$.

The same applies to points c and d, when the roles of transmitter 1 and transmitter 2 are exchanged.

A description of the gaussian case, as mentioned earlier, follows immediately from the theorems given for the general case, by interpreting these for continuous values. The signal that reaches the receivers is equal to $\mathbf{y} = \mathbf{x} + \mathbf{x}' + \mathbf{n}$, in which $\mathbf{n}$ represents the gaussian noise of power P_n. Consequently, the term $I(X;(Y/X'))$ of Theorem 8.3 can be written as

$$
\begin{aligned}
I(X;Y/X') &= H(Y/X') - H(Y/X,X') \\
&= H(X + X' + N/X') - H(X + X' + N/X,X') \\
&= H(X + N/X') - H(N/X,X').
\end{aligned}
\tag{8.36}
$$

Since the noise is entirely independent of X and X', it must follow that

$$
H(N/X,X') = H(N).
\tag{8.37}
$$

In addition, the mutual independence of X and X' will imply that

$$
H(X + N/X') = H(X + N).
\tag{8.38}
$$

This results in

$$
\begin{aligned}
I(X;Y/X') &= H(X + N) - H(N) \\
&= H(X + N) - \log \sqrt{2\pi e P_n} \\
&\leq \log \sqrt{2\pi e(P + P_n)} - \log \sqrt{2\pi e P_n},
\end{aligned}
\tag{8.39}
$$

in which P is the power constraint on transmitter 1.

This last expression reflects the fact that the measure of information of a gaussian distribution is always equal to the maximum value possible.

Rewriting the right-hand term leads to the simplification

$$
I(X;Y/X') \leq \tfrac{1}{2} \log \left(1 + \frac{P}{P_n}\right),
$$

and finally to

$$
R_1 \leq \tfrac{1}{2} \log \left(1 + \frac{P}{P_n}\right).
\tag{8.40}
$$

In the same manner, we can derive expressions for the remaining constraints of Theorem 8.3:

$$R_2 \le \frac{1}{2}\log\left(1 + \frac{P'}{P_n}\right) \tag{8.41}$$

with P' equal to the power constraint on transmitter 2 and

$$R_1 + R_2 \le \frac{1}{2}\log\left(1 + \frac{P + P'}{P_n}\right). \tag{8.42}$$

The above calculations are all expressed in terms of bits per transmission. However, continuous signals, which last for a period T, can be represented by $2WT$ samples, with W equal to the bandwidth of the signal, so in effect, $2W$ samples are taken per second. By multiplying the right parts of formulae (8.40)–(8.42) by $2W$, we will obtain upper limits for the rates, stated in bits/sec. This corresponds to formulae (8.3) and (8.4).

The equality of expressions (8.42) and (8.44) applies when X and X' are characterised by gaussian distributions.

These three expressions enable us to define the capacity region of Figure 8.4 for a gaussian multi-access channel. If transmitter 1 were to send no information at all, i.e. $R_1 = 0$, then the maximum rate for transmitter 2 would be equal to $\frac{1}{2}\log(1 + P'/P_n)$. Refer to point a of Figure 8.5. Vice versa for point d of the figure. The corners b and c can be found as follows. First, the receiver decodes the message sent by transmitter 3 by regarding the information sent by transmitter 1 as noise. The rate which still permits an arbitrarily small error probability is equal to

$$R_2 \le \frac{1}{2}\log\left\{1 + \frac{P'}{P + P_n}\right\}. \tag{8.43}$$

Figure 8.5. Capacity region for a gaussian multi-access channel with two transmitters.

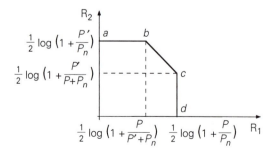

The decoded message of transmitter 2 can then be subtracted from the received message, after which the message of transmitter 1 can be decoded. This is a feasible method, provided

$$R_1 \le \frac{1}{2} \left\{ 1 + \frac{P}{P_n} \right\}. \tag{8.44}$$

This is exactly equal to the co-ordinates of corner c. The co-ordinates of point b can be found in the same manner.

8.3 Broadcast channels

A broadcast channel consists of one transmitter and several receivers. An example of this type of channel is given by a TV or radio transmitter, which broadcasts programmes that are received by a large number of receivers. Due to the inherent noise, no two receivers will actually receive identical signals. Consider Figure 8.6, which depicts a broadcast channel with k receivers.

In the subsequent discussion, we will assume that there are only two receivers: $k = 2$.

The transmitter will transmit messages X^n of length n, which are related to the original messages U^N of length N. Taking into account the noise present in the channel, decoder 1 will receive Y_1^n instead of X^n.

After this message has been decoded, the receiver will find V_1^N, which is a distorted version of U^N. In the same manner, decoder 2 will receive Y_2^n and decode this into V_2^N.

A general description of the channel can be given, in fact, by the transformation probabilities $p(y_1,y_2/x)$ from the alphabet of the encoder output to the channel input.

Figure 8.6. Broadcast channel.

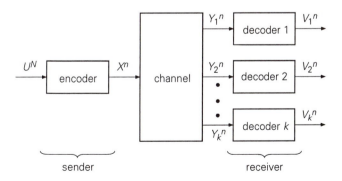

A memoryless *degraded broadcast channel* is a broadcast channel for which it is the case that

$$p(y_1, y_2/x) = p(y_1/x)\, p(y_2/x). \tag{8.45}$$

The *gaussian broadcast channel* is an example of a degraded broadcast channel. We can write $\mathbf{y}_1 = \mathbf{x} + \mathbf{n}_1$, with $\mathbf{n}_1$ representing gaussian noise of power P_{n_1}, and naturally also $\mathbf{y}_2 = \mathbf{x} + \mathbf{n}_2$. This expression can be rewritten as

$$\mathbf{y}_2 = \mathbf{x} + \mathbf{n}_2 = \mathbf{y}_1 + \mathbf{n}_3,$$

in which $\mathbf{n}_3$ represents a gaussian random variable with a power $P_{n_2} - P_{n_1}$. A schematic representation of this expression is given in Figure 8.7.
Clearly, the broadcast channel can be considered as a number of receivers that are connected in series and that transmit messages to each other. The transmitted messages degrade a little with each step.

Considering the configuration as given in Figure 8.6 with $k = 2$, then the gaussian case will yield the following equations for the individual capacities:

$$C_1 = W \log \left\{ 1 + \frac{P_x}{P_{n_1}} \right\} \tag{8.46}$$

and

$$C_2 = W \log \left\{ 1 + \frac{P_x}{P_{n_2}} \right\}, \tag{8.47}$$

in which P_x is the input power of the channel.
In the subsequent discussion we will assume that $P_{N_1} < P_{N_2}$ and therefore $C_1 > C_2$.

Obviously, time sharing offers a practical solution. In this case the channel is first made available for the transmission of messages to decoder 1 and subsequently to decoder 2, respectively. The corresponding time sharing rates are plotted in Figure 8.8.

Figure 8.7. The degraded broadcast channel.

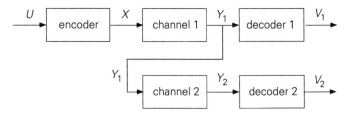

However, there are methods of improving the performance of the channel. Assume a given message $\mathbf{x}$ consists of the superposition of two messages $\mathbf{x}_1$ and $\mathbf{x}_2$, with related powers αP_x and $(1 - \alpha)P_x$ respectively where $0 \le \alpha \le 1$, where $\mathbf{x}_2$ is intended to be received by the high noise receiver 2. Receivers 1 and 2 will thus receive $\mathbf{y}_1 = \mathbf{x}_1 + \mathbf{x}_2 + \mathbf{n}_1$ and $\mathbf{y}_2 = \mathbf{x}_1 + \mathbf{x}_2 + \mathbf{n}_2$, respectively. See Figure 8.9.

Receiver 2 will interpret P_{n_2} and αP_x as noise sources, whose total power is equal to $\alpha P_x + P_{n_2}$. Consequently, messages can be sent to receiver 2 with a sufficiently small error probability provided the transmission rate is smaller than

$$C_2(\alpha) = W \log \left\{ 1 + \frac{(1 - \alpha)P_x}{\alpha P_x + P_{n_2}} \right\}. \tag{8.48}$$

Figure 8.8. The capacity region for the case of time sharing.

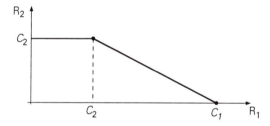

Figure 8.9.

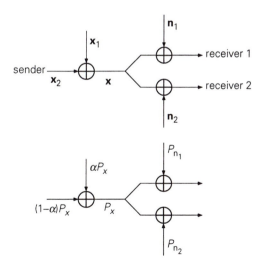

Since $P_{n_1} < P_{n_2}$, receiver 1 will also be able to receive message $\mathbf{x}_2$ with an error probability which is sufficiently small.

After decoding $\mathbf{x}_2$, receiver 2 can subtract $\mathbf{x}_2$ from Y_1, resulting in

$$\mathbf{y}_1 - \mathbf{x}_2 = \mathbf{x}_1 + \mathbf{n}_1.$$

Now receiver 1 has to communicate via a channel with additive gaussian noise and an input power αP_x. The capacity of this channel is

$$C_1(\alpha) = W \log \left(1 + \frac{\alpha P_x}{P_{n_1}}\right). \tag{8.49}$$

Therefore, receiver 1 is capable of receiving $\mathbf{x}_1$ as well as $\mathbf{x}_2$ correctly. This implies that rates

$$R_1 = W \log \left\{ 1 + \frac{(1 - \alpha P_x)}{\alpha P_x + P_{n_2}} \right\} + W \log \left(1 + \frac{\alpha P_x}{P_{n_1}}\right), \tag{8.50}$$

$$R_2 = W \log \left\{ 1 + \frac{(1 - \alpha)P_x}{\alpha P_x + P_{n_2}} \right\} \tag{8.51}$$

can be obtained.

These rates are better than those obtained by time sharing, as is illustrated in Figure 8.10.

What actually happens here is described by the following. The high noise receiver 2 gets a degenerated version of $\mathbf{x}$, whereas the low noise receiver 1 also receives additional information, which enables it to still obtain the original $\mathbf{x}$. For instance, this situation occurs with TV stations which broadcast programmes in High Definition TV (HDTV). The programmes are received by ordinary TV sets as well as special HDTV sets for high image quality. Using the same terminology as the above discussion, we can say

Figure 8.10. The capacity region for a gaussian broadcast channel with two receivers.

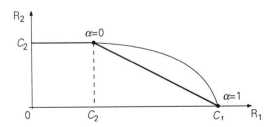

that ordinary TV sets can decode the signal x_2 for producing ordinary TV pictures. In addition to x_2, HDTV sets are also capable of decoding x_1, thus achieving a higher picture quality.

We will now attempt to describe the capacity region of a more general case of a degraded broadcast channel. We will examine the discrete case and start by restating Fano's inequality, as given earlier, for coded messages that are transmitted through a broadcast channel. The resulting theorem will play an important role in our analysis of the degraded broadcast channel.

Theorem 8.4
Assume an encoder is given which can transcribe messages U^N of length N to coded words X^n of length n, which form the input of a channel. The output of the channel is denoted by Y^n, whereas the messages V^N are found after decoding.
For the error probability at the output of the channel it is the case that

$$P_e \log (J - 1) + H(P_e) \geq \frac{1}{N} H(X^n/Y^n), \tag{8.52}$$

in which J represents the magnitude of the input alphabet.

Proof
By using Fano's inequality (refer to Theorem 4.4), we can write for the ith symbol of the encoder input and that of the decoder output

$$P_e \log (J - 1) + H(P_e) \geq H(U_l/V_l). \tag{8.53}$$

By calculating the mean values of both sides of formula (8.53) for all values of l and taking into account the fact that the information measures and the error probabilities are concave functions, we find

$$P_e \log (J - 1) + H(P_e) \geq \frac{1}{N} \sum_{l=1}^{N} H(U_l/V_l). \tag{8.54}$$

It is also the case that

$$H(U^N/V^N) = H(U_1/V^N) + H(U_2/V^N,U_1) + \ldots$$
$$+ H(U_n/V^N,U_1,\ldots,U_{N-1})$$

Figure 8.11. The communication scheme with coding.

$$\leq \sum_{l=1}^{N} H(U_l/V_l). \tag{8.55}$$

Therefore, with the aid of equation (8.54), we find

$$P_e \log (J-1) + H(P_e) \geq \frac{1}{N} H(U^N/V^N). \tag{8.56}$$

Finally, since there is a one-to-one relation between U^N and X^n, we can use Theorem 8.1 to write

$$P_e \log (j-1) + H(P_e) \geq \frac{1}{N} H(U^N/V^N)$$

$$= \frac{1}{N} \{ H(U^N) - I(U^N;V^N) \}$$

$$\geq \frac{1}{N} \{ H(X^n) - I(X^n;Y^n) \}$$

$$= \frac{1}{N} H(X^n/Y^n), \tag{8.57}$$

arriving at the expression we wished to prove. $\qquad \square$

Figure 8.12 is an expanded version of Figure 8.7, which is used for explaining the following theorem.

Here, we will assume that the signal which is to be transmitted consists of the superposition of two components: U^N and U'^N, where U'^N (and therefore also W^m) represents the information meant for both receivers and where U^N represents the extra information for receiver 1. Think, for instance, of the earlier example of HDTV. The index N denotes the length of the coded and decoded messages; n is the length of the code words which form the input and output of the channels.

Figure 8.12. The degraded broadcast channel.

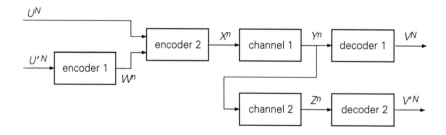

If the situation is regarded from the point of view of receiver 2, then the sequence of encoder 2, channel 1 and channel 2 can be considered as one single new channel. This results in Figure 8.13.

Theorem 8.5
Consider a degraded broadcast channel. Provided that for every auxiliary random variable W and channel input X it is the case that

$$R_1 > I(X;Y/W) \tag{8.58}$$

or

$$R_2 > I(W;Z), \tag{8.59}$$

then there will be no codes with rates (R_1,R_2) which allow the transfer of information to occur without any errors.

Proof
For rate R_2 it is the case that

$$R_2 = \frac{1}{n} H(U'^N) = \frac{1}{n} \left\{ I(U'^N;Z^n) + H(U'^N/Z^n) \right\}. \tag{8.60}$$

The following expressions can be derived for $I(U'^N;Z^n)$:

$$I(U'^N;Z^n) = H(Z^n) - H(Z^n/U'^N)$$

$$\leq \sum_{l=1}^{n} H(Z_l) - \sum_{l=1}^{n} H(Z_l/U'^N,Z_{l-1},...,Z_1). \tag{8.61}$$

Clearly, it is the case for all l that

$$H(Z_l/U'^N,Z_{l-1},...,Z_1) \geq H(Z_l/U'^N,Z_{l-1},...,Z_1,Y_{l-1},...,Y_1)$$

and therefore

Figure 8.13. Alternative scheme of the degraded broadcast channel.

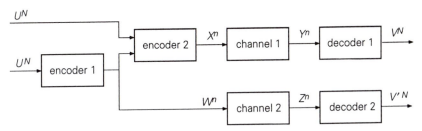

$$I(U'^N;Z^n) \le \sum_{l=1}^{n} \{ H(Z_l) - H(Z_l/U'^N, Z_{l-1}, \ldots, Z_1, Y_{l-1}, \ldots, Y_1) \}. \quad (8.62)$$

If U'^N and $Y_{l-1}, \ldots, Y_1$ are given then Z_l is conditionally independent of $Z_{l-1}, \ldots, Z_1$.
It follows that

$$H(Z_l/U'^N, Z_{l-1}, \ldots, Z_1, Y_{l-1}, \ldots, Y_1) = H(Z_l/U'^N, Y_{l-1}, \ldots, Y_1)$$

and thus

$$I(U'^N;Z^n) \le \sum_{l=1}^{n} I(Z_l;(U'^N, Y_{l-1}, \ldots, Y_1)). \quad (8.63)$$

By replacing $(U'^N, Y_{l-1}, \ldots, Y_1)$ by a new discrete random variable W_l this expression can be rewritten in the more simple form

$$\frac{1}{n} I(U'^N;Z^n) \le \frac{1}{n} \sum_{l=1}^{n} I(Z_l;W_l). \quad (8.64)$$

The right-hand side of this expression can be regarded as the expectation of the mutual information $I(Z;W)$ per symbol.
Thus, equation (8.64) yields

$$\frac{1}{n} I(U'^N;Z^n) \le I(Z;W)$$

With the aid of equation (8.60), we can find for R_2

$$R_2 = \frac{1}{n} \{ I(U'^N;Z^n) + H(U'^N/Z^n) \}$$

$$\le I(Z;W) + \frac{1}{n} H(U'^N/Z^n). \quad (8.65)$$

By introducing Fano's inequality at this point, as given in Theorem 8.4, we find that

$$P_e \log (J' - 1) + H(P_e) \ge \frac{1}{N} H(U'^N/Z^n)$$

$$\ge \frac{n}{N} (R_2 - I(Z;W)) \quad (8.66)$$

in which J' is the magnitude of the encoder input.

As long as $R_2 > I(Z;W)$, the error probability can no longer decrease indefinitely. Therefore, the second half of the theorem is demonstrated.

The first half of the theorem can be demonstrated in the same manner. Remembering Theorem 8.1, we can write for the rate R_1

$$R_1 = \frac{1}{n} H(U^N/U'^N) = \frac{1}{n} \{ I(U^N;Y^n/U'^N) + H(U^N/Y^n,U'^N) \}$$

$$\leq \frac{1}{n} \{ I(X^n;Y^n/U'^N) + H(U^N/Y^n) \}. \tag{8.67}$$

By definition, it is the case that

$$I(X^n;Y^n/U'^N) = H(Y^n/U'^N) - H(Y^n/U'^N,X^n)$$

$$= \sum_{l=1}^{n} H(Y_l/U'^N,Y_{l-1},\ldots,Y_1)$$

$$- \sum_{l=1}^{n} H(Y_l/U'^N,X^n,Y_{l-1},\ldots,Y_1). \tag{8.68}$$

Since the channel is memoryless, Y_l with respect to X^n will only depend on X_l and therefore

$$H(Y_l/U'^n,X^n,Y_{l-1},\ldots,Y_1) = H(Y_l/U'^N,X_l,Y_{l-1},\ldots,Y_1).$$

Furthermore, with equation (8.68) this leads to

$$\frac{1}{n} I(X^n;Y^n/U'^N) = \frac{1}{n} \sum_{l=1}^{n} H(Y_l/U'^N,Y_{l-1},\ldots,Y_1)$$

$$- \frac{1}{n} \sum_{l=1}^{n} H(Y_l/U'^N,X_l,Y_{l-1},\ldots,Y_1)$$

$$= \frac{1}{n} \sum_{l=1}^{n} I(X_l;Y_l/U'^N,Y_{l-1},\ldots,Y_1)$$

$$= I(X;Y/W). \tag{8.69}$$

The final step follows by introducing a new random variable W and realising the term containing the summation can be regarded as an expectation.

Combining these last observations, we find

$$H(U^N/Y^n) \geq n\mathsf{R}_1 - I(X^n;Y^n/U'^N)$$

$$= n\{\mathsf{R}_1 - I(X;Y/W)\}. \tag{8.70}$$

Again, we can use Fano's inequality, though now the lower limit is given in terms of $H(U^N/Y^n)$.

If the error probability is to be chosen arbitrarily small, then it must never be the case that $R_1 > I(X;Y/W)$.

It will only be possible to find a suitable code, which allows information to be transmitted with an arbitrarily small error probability, if both conditions stated by the theorem are satisfied. □

The converse theorem can also be demonstrated. However, we will not give a full proof, since the theorem becomes self-evident on considering Theorem 8.5.

Theorem 8.6
If the joint probability distribution of the input X and auxiliary random variable W are such that

$$\mathsf{R}_1 \leq I(X;Y/W) \tag{8.71}$$

and

$$\mathsf{R}_2 \leq I(W;Z), \tag{8.72}$$

then theoretically, a code with rate (R_1,R_2) can be found which will allow for an arbitrarily small error probability. □

It is obvious that the capacity region of memoryless degraded broadcast channels consists of the convex hull of all R_1 and R_2 which satisfy the conditions stated in the previous theorem, for every W jointly distributed with X.

Figure 8.14. Broadcast channel with two receivers.

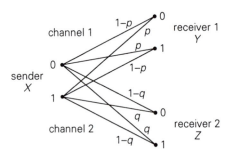

We will now provide an example of the capacity region for symmetric binary channels.

Figure 8.12 illustrated the fact that the channel for receiver 2 can be described by the series connection of the channel for receiver 1 and another channel, which together are completely equivalent to the original channel for receiver 2. Channel 2 of Figure 8.14 can be regarded as channel 1, followed by another channel, as depicted in Figure 8.15.

Figure 8.15.

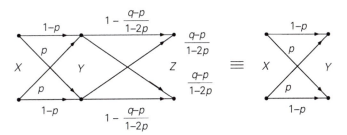

Figure 8.16.

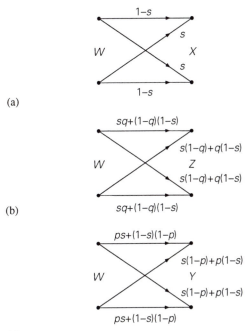

(a)

(b)

(c)

It can be confirmed without difficulty that the transition probabilities of this cascade are equal to q and $1 - q$, respectively.

In order to determine the capacity region, based on Theorem 8.6, we will introduce an auxiliary variable W, which is connected to X by a binary symmetric channel with transition probability s (see Figure 8.16(a)). Corresponding to Theorem 8.6, we are interested in finding $I(W;Z)$ and $I(X;(Y/W))$. Let the probabilities pertaining to W be denoted by α and $1 - \alpha$, respectively. When considering the symmetry of the interconnections, it becomes evident that the rates will have their maximum when $\alpha = \frac{1}{2}$. With the aid of Figure 8.16(b), we will then find that

$$\begin{aligned} I(W;Z) &= H(Z) - H(Z/W) \\ &= 1 + t \log t + (1 - t) \log(1 - t), \end{aligned} \tag{8.73}$$

with $t = sq + (1 - q)(1 - s)$.

For $I(X;(Y/W))$ we can write by using Figure 8.16(c)

$$\begin{aligned} I(X;Y/W) &= H(Y/W) - H(Y/W,X) \\ &= H(Y/W) - H(Y/X) \\ &= -v \log v - (1 - v) \log(1 - v) + p \log p \\ &\quad + (1 - p) \log (1 - p) \end{aligned} \tag{8.74}$$

with $v = sp + (1 - p)(1 - s)$.

By varying the value of s, we can calculate a number of values which define the boundaries of the capacity region. For instance, when $s = 0$, it follows that $R_1 = 0$, since $I(X;(Y/W)) = 0$ and $R_2 = 1 + q \log q + (1 - q) \log(1 - q) = 1 - H(Q)$. When $s = \frac{1}{2}$ our calculations result in $R_2 = 0$ and $R_2 = 1 + p \log p + (1 - p) \log(1 - p) = 1 - H(P)$. The capacity region is given in Figure 8.17.

8.4 Two-way channels

A two-way channel will always comprise at least two parties which can both

Figure 8.17. The capacity region of a binary symmetric broadcast channel.

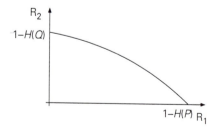

act as transmitter and receiver. Obviously, this situation can occur if each party performs a single role as either transmitter or receiver and by repeatedly exchanging the roles of the parties. However, it is typical of a two-way channel that both parties transmit their messages to each other simultaneously and that a message in one direction will interfere with a message in the other direction. Figure 8.18 shows a general diagram of a two-way channel.

One of the most well-known examples of a two-way channel is the *binary multiplying channel*, with binary input and output alphabets. On every clock pulse, each user will transmit their bit of information through the channel and simultaneously receive a code bit, which is in fact the logical 'AND' of the transmitted bits. If the information bit transmitted by a user is equal to 1 and the bit received from the channel is also a 1, then the user can conclude that the other party must also have transmitted a 1.

However, if the bit received is a 0, then the other user must have sent a 0. If the first user sent a 0 instead, then he will not be capable of determining whether the other user sent a 0 or a 1. Luckily, though, the Hagelberger code can offer a solution to this problem. (See Figure 8.19.)

Each encoder can assume one of two states, either state 1 or state 2. The current state of an encoder will depend on the code bit received last and on the state during the previous clock pulse. Therefore, both encoders will always be in identical states. An encoder can only enter state 1 if either the received bit is a 0, or the previous state was state 2. Consequently, an

Figure 8.18. Two-way channel.

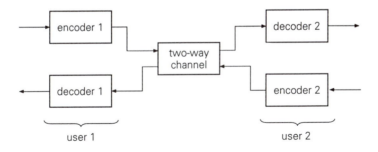

Figure 8.19. Hagelberger code.

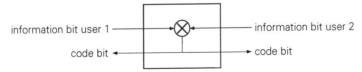

encoder can only enter state 2 if the bit received last was equal to 1 and the encoder was not in state 2 already. In state 1 the encoders will transmit the next data bit. In state 2 they will transmit the complement of the previous data bit.

Figure 8.20 shows the resulting code bits as a function of the data bits. The number of code bits required to transfer a single bit will depend on the data bits transmitted by both users and with be equal to 1 or 2. If the probability of the data bits sent by the users is given by $p(1) = p$, then for each data bit the total number of code bits will be equal to

$$L = 1 \cdot p^2 + 2p(1-p) + 2(1-p)^2$$

$$= 2 - p^2. \tag{8.75}$$

Therefore, the code rate in either direction will amount to

$$R = \frac{H(P)}{2 - p^2}, \tag{8.76}$$

where $H(P) = -p \log p + (1-p) \log(1-p)$.
If $p = \frac{1}{2}$, then we find that

$$R = \frac{1}{7/4} \approx 0.571.$$

However, the maximum rate is achieved when $p = 0.63$. This rate is equal to $R = 0.593$. Therefore, if the users wish to transmit information for which $p = 0.5$, then in order to achieve a maximum transmission rate, before the information is passed to the encoder, it must first be transformed (precoded), to ensure that the probability of a 1 is equal to 0.63. This task can be performed by a Huffmann decoder. Since a Huffmann code is used to transform the original data, with different probabilities per symbol, to code words with symbols of equal probability, obviously, the reverse operation

Figure 8.20.

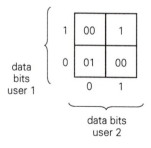

will achieve the desired effect; the symbols will be transformed from equally probable symbols to symbols with differing probabilities.

The receiver will then apply a Huffman code. In addition to the Hagelberger code, the Schalkwijk code can also be used, which will yield a rate of 0.619 bits per clock interval, simultaneously in both directions.

It is not possible to give a general expression for the capacity of a memoryless two-way channel. However, we can still calculate the values of the inner and the outer bounds of the capacity.

Let X and X' be random variables on the channel input alphabets. Y and Y' are the random variables on the channel output alphabets.

The following theorem provides an outer bound for the capacity region.

Theorem 8.7
Let (X,X',Y,Y') have joint probability distributions $p(x_l,x_l')$ and $p(y_l,y_l'/x_l,x_l')$ for each joint probability distribution $p(x_l,x_l')$. Then the capacity region of the memoryless two-way channel (see Figure 8.21) will be enclosed by the boundaries for which

$$(\mathsf{R}_1,\mathsf{R}_2): \quad \mathsf{R}_1 \le I(X;Y'/X') \text{ and } \mathsf{R}_2 \le I(X';Y/X). \tag{8.77}$$

Proof
We must prove that for the pairs $(\mathsf{R}_1,\mathsf{R}_2)$ which do not satisfy the conditions of the theorem, it is not possible to find codes which allow the error probability to be reduced to an arbitrarily small value. Obviously, it is the case that

$$H(U^N) = I(U^N;Y'^n) + H(U^N/Y'^n) \tag{8.78}$$

and

$$H(U^N/U'^N) = I(U^N;Y'^n/U'^N) + H(U^N/Y'^n,U'^N). \tag{8.79}$$

Since there is a one-to-one correspondence on the one hand between U^N and

Figure 8.21. Two-way channel.

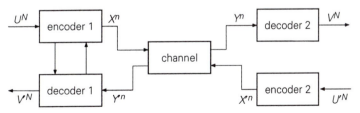

X^n and on the other hand one between U'^N and X'^n, it is the case that

$$H(U^N/U'^N) = I(X^n;Y'^n/X'^n) + H(X'^n/Y'^n,X'^n).$$

Furthermore, it is true that

$$I(X^n;Y'^n/X'^n) \leq \sum_{l=1}^{n} I(x_l;y'_l/X'^n)$$

$$= \sum_{l=1}^{n} \left\{ \sum_{x_l,X'^n,y'_l} p(x_l,X'^n,y'_l) \log \left\{ \frac{p(x_l/N'^n,y'_l)}{p(x_l/X'^n)} \right\} \right\}$$

$$= \sum_{l=1}^{n} \left\{ \sum_{x_l,X'^n,y'_l} p(x_l,X'^n,y'_l) \log \left\{ \frac{p(x_l,X'^n,y'_l)/p(X'^n,y'_l)}{p(x_l,X'^n)/p(X'^n)} \right\} \right\}$$

$$= \sum_{l=1}^{n} \left\{ \sum_{x_l,X'^n,y'_l} p(x_l,X'^n,y'_l) \log \left\{ \frac{p(y'_l/x_l,X'^n)p(x_l,X'^n)}{p(y'_l/X'^n)p(X'^n)p(x_l,X'^n)/p(X'^n)} \right\} \right\}$$

$$= \sum_{l=1}^{n} \left\{ H(Y'_l/X'^n) - H(Y'_l/X_l,X'^n) \right\}$$

$$\leq \sum_{l=1}^{n} H(Y_l/X'_l) - \sum_{l=1}^{n} H(Y'_l/(X_l,X'_l,X'_{l-1},\ldots,X'_1)); \tag{8.80}$$

since we have a memoryless channel

$$H(Y'_l/(X_l,X'_l,X'_{l-1},\ldots,X'_1)) = H(Y'_l/X_l,X'_l),$$

and thus, equation (8.80) will yield

$$I(X^n;Y'^n/X'^n) \leq \sum_{l=1}^{n} H(Y'_l/X'_l) - \sum_{l=1}^{n} H(Y'_l/X_l,X'_l)$$

$$= \sum_{l=1}^{n} I(X_l;Y'_l/X'^n)$$

$$= nI((X;Y'/X'). \tag{8.81}$$

For the rate per symbol R_1, it is the case that

$$R_1 = \frac{1}{n} H(U^N/U'^N) = \frac{1}{n} \left\{ I(X^n;Y'^n/X'^n) + H(X'^n/Y'^n,X'^n) \right\}$$

$$\leq I(X/Y'/X') + \frac{1}{n} H(X^n/Y'^n, X'^n) \qquad (8.82)$$

and therefore

$$H(X^n/Y'^n, X'^n) \geq n \{R_1 - I(X;Y'/X')\}. \qquad (8.83)$$

With the aid of Fano's inequality, it should be the case that

$$n \{P_e \log (J - 1) + H(P_e)\} \geq H(X^n/Y'^n, X'^n) \geq n \{R_1 - I(X;Y'/X')\}. \qquad (8.84)$$

If the error probability is to approach zero, then it must be the case that

$$R_1 \leq I(X;Y'/X'). \qquad (8.85)$$

Since the roles of X and X' are symmetrical in these expressions, it is evident that for R_2 it must be true that

$$R_2 \leq i(X';Y/X).$$

This is the expression we wished to prove. □

In the above theorem, we assumed that the encoders have some means of co-operating with each other. This assumption was not expressed explicitly, but implied by the use of the joint probability distributions $p(x_l, x'_l)$ instead of $p(x_l) \cdot p(x'_l)$.

However, in practice, an encoder will only have partial knowledge of the code word the other encoder wishes to send. Upon receiving the first symbol, the uncertainty will be the greatest, but towards the end of the transmitted block, the certainty with respect to the message will increase. This is the reason that the limit stated by the theorem is an outer bound. An inner bound can be found by assuming that the encoders operate entirely independently of each other: $p(x_l, x'_l) = p(x_l) \cdot p(x'_l)$. For all (X, X', Y, Y') with the corresponding joint probability distribution $p(x_l)p(x'_l)p(y'_l, y'_l/x_l, x'_l)$, it holds that the convex hull of the set

$$\{(R_1, R_2) \mid R_1 \leq I((X/X');Y) \text{ and } R_2 \leq i((X'/X);Y)\}$$

is enclosed within the capacity region of the two-way channel.

The inner and outer bounds are given in Figure 8.22.

Figure 8.22. Inner and outer bounds for the two-way channel.

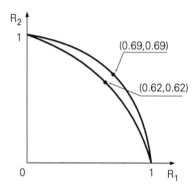

8.5 Exercises

8.1. Consider a multi-access channel with two transmitters and with binary inputs. The output is given by

$$Y = X_1 * X_2$$

in which $*$ denotes some undefined operator.

a. If the operator $*$ represents a multiplication, we will have a binary multiplier channel (BMC). Determine the capacity region for this channel.

b. The binary erasure multi-access channel (BEMC) will result, in the case where the operator $*$ represents an addition. The output is assumed to be ternary. Determine the capacity region for the binary multiplier channel.

8.2. Consider a multi-access channel with k users in which additive gaussian noise causes interference. Assume that each user transmits the same signal power level P_x. Then for each user the signal-to-noise ratio is equal to $P_x/P_n = 10$.

a. Calculate the capacity of the channel for the case in which frequency-division is applied and each of the k users is assigned an equal bandwidth w.

b. Plot the capacity per unit bandwidth for the case described by a as a function of k.

c. Give an expression for the capacity available to each user, if time sharing is applied, instead of frequency-division, assuming that each user may access the channel for the same length of time.

8.3. Two transmitters can access a gaussian multi-access channel. Transmitter 1 has a signal-to-noise ratio of 20, transmitter 2 has a signal-to-noise ratio of 10.

a. Calculate the capacity in bits per transmission of transmitter 1, assuming transmitter 1 has exclusive access to the channel. Repeat this calculation for transmitter 2.

b. Give the capacity region of the gaussian multi-access channel, for the case in which both transmitters are using the channel simultaneously.

8.4. Consider a broadcast channel with two receivers. The channel between the transmitter and receiver 1 can be regarded as a binary symmetrical channel, for which the probability of an incorrect transmission is equal to $p = 0.1$. For the channel between the transmitter and receiver 2, the probability of an incorrect transmission is equal to $q = 0.6$.

a. Calculate the capacity of the channel between the transmitter and receiver 1. Also calculate the channel capacity between the transmitter and receiver 2.

b. Find two expressions in terms of the mutual information, which together define the capacity region for the broadcast channel considered here.

c. Plot the capacity region of this broadcast channel.

8.6 Solutions

8.1. *a.* If $X_1 = 1$, then transmitter 2 will be able to send information to the receiver with a maximum rate of 1 bit per transmission. The same holds for transmitter 1. When $X_2 = 1$, transmitter 1 will also send information with a rate of 1 bit to the receiver. Therefore, two points are defined on the R_2 and R_1 axes, respectively. Time sharing will allow any combination of rates to

Figure 8.23. Capacity regions for BMC and BEMC.

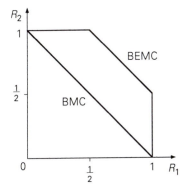

be obtained for which $R_1 + R_2 = 1$. The corresponding capacity region is given in Figure 8.23.

b. If the output Y assumes the value either 0 or 2, then there is no uncertainty as to the information bits sent by the two transmitters. There is uncertainty, however, when Y is equal to 1.

If $X_1 = 0$, then transmitter 2 can send information to the receiver with a rate of 1 bit per transmission: $R_2 = 1$. Naturally, R_1 will be equal to 1 when $X_2 = 0$. Therefore, two points of the outer boundary of the capacity region are defined. If $R_1 = 1$, then the bits sent by transmitter 1 can be regarded as noise by transmitter 2. Half of the bits from transmitter 1 that are added to the information bits of transmitter 2 will be zeros, whereas the other half of the bits will be ones. For transmitter 2, the multi-access channel can be described in the same manner as a binary erasure channel (BEC), with transition probability (1,2). In Exercise 4.7 we already found that the capacity of the binary erasure channel is equal to $C = 1 - p$. Therefore, here we will find that $C = \frac{1}{2}$.

This means that when transmitter 1 is sending information with a rate $R_1 = 1$, transmitter 2 will still be capable of sending $\frac{1}{2}$ a bit per transmission to the receiver. Conversely, if for transmitter 2, $R_2 = 1$, then transmitter 1 will also be able to send $\frac{1}{2}$ a bit. The resulting capacity region is given in Figure 8.23.

8.2. *a.* In the case of frequency division multi-access, it is the case that

$$C^* = \frac{W}{k} \log \left\{ 1 + \frac{kP_x}{P_n} \right\} = \frac{W}{k} \log \left\{ 1 + 10k \right\}.$$

b. The following table gives the values of C/W corresponding to several values of k.

k	C/W
1	3.46
2	2.20
3	1.65
4	1.40
5	1.13
10	0.66
20	0.38
30	0.27
40	0.22
50	0.18
100	0.13

For large values of k we find that $C/W \approx \frac{1}{k} \log 10k$. The curve corresponding to C/W is plotted in Figure 8.24.

Figure 8.24. Capacity of the multi-access channel of Exercise 8.2.

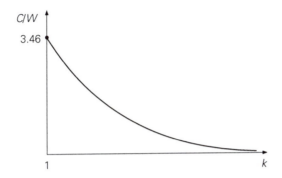

c. Since time sharing effectively results in the same capacity as frequency division, we will find the same results as for *a* and *b*.

8.3. *a*. The capacity is given by

$$C = W \log \left(1 + \frac{P_x}{P_n}\right) \text{ bits/sec}$$

or

$$C = \frac{1}{2} \log \left(1 + \frac{P_x}{P_n}\right) \text{ bits/transmission.}$$

Therefore, for transmitter 1 and 2 it follows that

$$C_1 = \frac{1}{2} \log \left(1 + \frac{P_1}{P_n}\right) = \frac{1}{2} \log 21 = 2.20$$

and

$$C_2 = \frac{1}{2} \log \left(1 + \frac{P_2}{P_n}\right) = \frac{1}{2} \log 11 = 1.73,$$

respectively.

b. The two values calculated for the previous solution *a* correspond to the edges of the capacity region.

The theory states that when transmitter 2 is sending information with a maximum rate, the maximum rate for transmitter 1 is equal to $\frac{1}{2} \log (1 + P_1/(P_2+P_n))$. After all, the signal sent by transmitter 2 may be regarded by transmitter 1 as an extra noise source.

Since $P_1 = 20P_n$ and $P_2 = 10P_n$, it follows that $P_1/(P_2 + P_n) = 20P_n/11P_n = 20/11$ and therefore

$$\frac{1}{2}\log\left(1 + \frac{P_1}{P_2 + P_n}\right) = \frac{1}{2}\log\left(1 + \frac{20}{11}\right) = \frac{1}{2}\log\frac{31}{11} = 0.75.$$

In the same manner, we can conclude that when transmitter 1 sends information with a maximum rate, the maximum rate for transmitter 2 is equal to

$$\frac{1}{2}\log\left(1 + \frac{P_2}{P_1 + P_n}\right) = \frac{1}{2}\log\left(1 + \frac{10}{21}\right) = \frac{1}{2}\log\frac{31}{21} = 0.28.$$

Therefore, the entire capacity region is given by Figure 8.25.

Figure 8.25. Capacity region of the gaussian multi-access channel of Exercise 8.3.

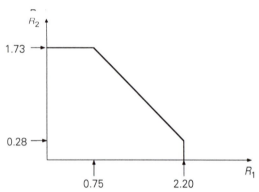

8.4. *a.* Since we are dealing with two binary symmetric channels, we can refer to Example 4.3, where we already calculated the capacity for the general case.
For receiver 1 it is the case that

$$C_1 = 1 - H(P) = 1 + 0.1 \log 0.1 + 0.9 \log 0.9 = 0.531.$$

The capacity for receiver 2 is given by

$$C_2 = 1 - H(Q) = 1 + 0.6 \log 0.6 + 0.4 \log 0.4 = 0.029.$$

b. The channel for receiver 2 is regarded as a cascade of two channels, i.e. the channel for receiver 1 and an additional channel, which ensures that the cascade has the same transition probabilities as the original channel.
The capacity region is implicitly given by $I(W;Z)$ and $I(X;(Y/W))$, in which X represents the channel input produced by the transmitter and Y and Z

represent the channel outputs for transmitter 1 and transmitter 2, respectively. W is an auxiliary variable.
The expression for $I(W;Z)$ is given by

$$I(W;Z) = H(Z) - H(Z/W) = 1 + t \log t + (1 - t) \log (1 - t)$$

for which

$$t = sq + (1 - q)(1 - s) = 0.6s + 0.4(1 - s) = 0.2s + 0.4$$

and s is equal to the transition probability, whose value is varied, assuming the channel between W and X is binary symmetrical.
For $I(X;(Y/W))$ we can write

$$\begin{aligned} I(X;(Y/W)) &= -v \log v - (1 - v) \log (1 - v) + p \log p \\ &\quad + (1 - p) \log (1 - p) \end{aligned}$$

$$\begin{aligned} &= -v \log v - (1 - v) \log (1 - v) + 0.1 \log 0.1 \\ &\quad + 0.9 \log 0.9 \end{aligned}$$

$$= -v \log v - (1 - v) \log (1 - v) - 0.469$$

with

$$v = sp + (1 - p)(1 - s) = 0.1s + 0.9(1 - s) = 0.9 - 0.8s.$$

c. No explicit expression exists for the capacity region. However, points of the region can be found by varying the value of s between 0 and 1.
If $s = 0$, it follows that $t = 0.4$ and $v = 0.9$ and, consequently,

$$I(W;Z) = 1 + 0.4 \log 0.4 + 0.6 \log 0.6 = 0.029,$$

$$I(X;(Y/W)) = -0.9 \log 0.9 - 0.1 \log 0.1 - 0.47 = 0.$$

If $s = \frac{1}{2}$, it follows that $t = 0.5$ and $v = 0.5$; therefore

$$I(W;Z) = 1 + 0.5 \log 0.5 + 0.5 \log 0.5 = 0,$$

$$\begin{aligned} I(X;(Y/W)) &= -0.5 \log 0.5 - 0.5 \log 0.5 - 0.47 = 1 - 0.47 = \\ &\quad 0.531. \end{aligned}$$

This corresponds entirely with the values found for the capacities at a, assuming that the other transmitter refrains from transmitting.
For $s = 0.1, 0.2, 0.3$ and 0.4 we can calculate the following values:

$I(W;Z)$	$I(X;(Y/W))$
0.019	0.210
0.010	0.357
0.005	0.455
0.001	0.512

See Figure 8.26.

Figure 8.26. Capacity region of the broadcast channel of Exercise 8.4.

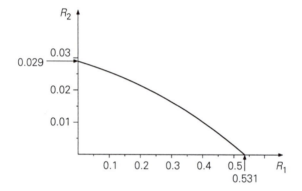

9

Error-correcting codes

9.1 Introduction

As was derived in Chapter 4, one can in principle realise error-free transmission through a discrete communication channel provided that the amount of information to be transported is at most C. This theorem gives no aid in constructing actual codes. It really only asserts the existence of such a code.

In this chapter we will examine the design of error-detecting and error-correcting codes in more detail. With respect to the study of these codes there are two fields of interest, namely the construction of codes on the one hand and their behaviour with regard to detection and correction of errors during the decoding of received words on the other hand. Decoding in particular is often complex in practice and thus requires the most attention. In this chapter we limit ourselves to simple linear block codes for the memoryless binary symmetric channel with an error probability p (see also Chapter 4).

The sequence of symbols generated by the information source is divided up into blocks of k symbols. We denote the message u consisting of k symbols by $u_1, u_2, \ldots, u_k$. We assume that these symbols originate from an alphabet that is identical to the channel alphabet. For a binary channel we therefore have $u_i \in \{0,1\}$, $i = 1, \ldots, k$. We furthermore assume that $p(u_i{=}0) = p(u_i{=}1) = \frac{1}{2}$, that is, we assume that we have utilised an optimal source code. As the channel is symmetric, this is also the optimal source probability distribution that makes certain that the channel capacity is achieved.

The number of possible messages u is therefore $M = 2^k$, and each message u has an equal probability $p(u) = 2^{-k}$.

When designing a channel code, one or more parity checks are added to these messages in an unambiguous manner. We denote the length of the M code words obtained in this manner by n. The code rate is now

$$R = \frac{\log M}{n}$$

$$= \frac{\log 2^k}{n} = \frac{k}{n} \text{ bits/symbol,} \tag{9.1}$$

and can be interpreted as a measure for the redundancy that is added to be able to detect or correct errors.

The construction of a code can therefore be perceived as choosing M channel code words from the in total 2^n possible code words. Decoding can now be regarded as deciding which channel word x was transmitted on the basis of a received (possibly distorted) word y.

If the received word y is not equal to a code word then at least one error has occurred. If y is equal to a code word then there may have been an errorless transmission, but this is by no means certain.

The $(n–k)$ parity checks can be used to add redundancy to the message so that one is able to detect or correct errors in the communication channel. The improvement of the error probability, however, is always at a cost to the amount of information transported per source symbol, R.

Consider so-called *repetition codes*. In these codes a symbol is repeated a number of times. Assume a source generates two symbols: 0 and 1. Let us assume that the error probability equals $p = 0.01$. In that case the probability of correct transmission of a symbol is 0.99. If we apply encoding such that code word 000 is assigned to symbol 0 and code word 111 to 1 (each generated source symbol is repeated 3 times) then the rate will decrease from $R = 1$ to $R = \frac{1}{3}$. However, the probability of correct receipt will increase. Correct receipt occurs if the code words are not affected by errors, or if just one symbol of a code word is changed. In the latter on the basis of a majority vote the correct input symbol will be found.

message	code word	received messages	reconstructed message
0	000	000 100 010 001	0
1	111	110 101 011 111	1

If there are 2 or 3 errors in the code words the wrong decision will be made.

The probability of error, P_e, now becomes

$$P_e = 1 - \{(0.99)^3 + 3 \cdot 0.01(0.99)^2\} \approx 3 \cdot 10^{-4}.$$

More generally in an $(n,1)$-repetition code a symbol is repeated $(n-1)$ times. Repetition of input symbols has thus a beneficial influence on the error probability.

Error correction is possible by counting the number of zeros in a received word and by deciding to go over to the symbol 0 if this number is greater than $n/2$. The code rate of this code is $R = 1/n$. In the following table, the result of this code for a few values of n is indicated for an error probability $p = 0.01$.

length n	remaining error probability	code rate
1	10^{-2}	1
3	$3 \cdot 10^{-4}$	1/3
5	10^{-5}	1/5
7	$4 \cdot 10^{-7}$	1/7
9	10^{-8}	1/9
11	$5 \cdot 10^{-10}$	1/11

The table shows that such a code has good error-correction properties, but that the code rate is low. If we let the number of repetitions, and thus the number of symbols per message, go to infinity, the error probability does indeed go to zero, but likewise the amount of transported information which makes this code unattractive in practice.

A correct approach can be achieved by constructing (n,k)-codes where we let both n and k grow but keep the number of parity checks, namely $n - k$, limited. Through this, a trade-off can be obtained between the number of errors to be corrected and the desired code rate.

9.2 Linear block codes

As has already been mentioned we have limited ourselves here to binary block codes. The *weight* $w(a)$ of a code word a is the number of ones in that code word. The *Hamming distance* $d(a,b)$ between two code words a and b is the number of positions where the code words have different symbols. For the code words

$$a = 1\ 0\ 1\ 0\ 1\ 0\ 1$$

$$b = 1\ 1\ 1\ 0\ 0\ 1\ 1$$

the weights are $w(a)$ = 4 and $w(b)$ = 5 respectively and their Hamming distance is $d(a,b)$ = 3.

When designing a code we can now choose the code words in such a way that the Hamming distance between each pair of code words has a certain minimum value. This minimum distance is called the *Hamming distance of the code*, denoted by d.

We can represent the code symbols using vectors in a linear vector space. The dimension of this space is equal to the number of symbols that make up a code word. For n = 3 we then find the representation shown in Figure 9.1 for the eight possible code words (000) through to (111) for the binary case. If the code now consists of just these eight code words then if an error occurs a transmitted code word would change into another code word and no error detection or correction would be possible. The Hamming distance is equal to 1 in this case. If we take a code with a Hamming distance 2 then the number of choosable code words is equal to 4, namely (000), (011), (101), (110). Now if exactly one error occurs it can be detected because the received word will differ then from the four code words. There are, however, still two possible code words that have a Hamming distance of 1 with respect to the received word. E.g. received code word (111) can come form either (011) or (101). So correction is not possible. Error correction

Figure 9.1. Three-dimensional representation of code words with length 3.

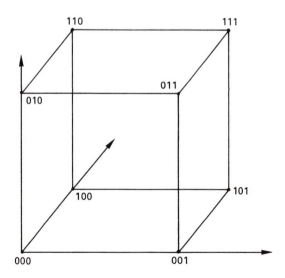

only becomes possible when $d = 3$, by only using the code words (000) and (111) for example. In this case if exactly one error occurs there is exactly one code word with the smallest Hamming distance, and it can then be concluded that this must be the transmitted code word.

It is apparent from the example that with a given number of messages $M = 2^k$ the length n of the channel words, and as such the number of parity checks $(n - k)$, is linked to the desired error-detection or error-correction performance of the code, and therefore to the necessary minimum Hamming distance of the channel words. In general, a code with Hamming distance d is capable of detecting $(d - 1)$ errors and of correcting $t = \lceil (d-1)/2 \rceil$ errors, where $\lceil a \rceil$ represents the largest integer less than a. Conversely, a code that must correct t errors must have a Hamming distance of at least $2t + 1$.

Example 9.1
Consider the following code with $n = 5$ and $k = 2$.

message	code word
(00)	(00000)
(10)	(10110)
(01)	(01101)
(11)	(11011)

Since the Hamming distance $d = 3$, this code can detect two errors and correct just one.   △

We speak of a *linear (binary) block code* C if the modulo 2 sum of every pair of code words is again a code word and if every code word when multiplied by a 0 or 1 also yields a code word. From these two requirements, it follows directly that the null vector is automatically a code word. Furthermore, it also follows that the minimum Hamming distance of a code C corresponds with the smallest weight of the code words other than the one related to the null vector:

$$d = \min \{ w(x) \mid x \in C, x \neq 0 \}, \tag{9.2}$$

where x represents a code word.

This can be seen as follows. For the Hamming distance of the code it is the case that

$$d = \min \{ d(a,b) \mid a,b \in C \text{ and } a \neq b \}$$

$$= \min \{ d(0,a + b) \mid a,b \in C \text{ and } a \neq b \}.$$

Since $a + b$ is also a code word of C, say x, and since $d(0,x) = w(x)$ equation (9.2) follows directly.

We will now consider how linear block codes can be designed. As an example a (7,4)-code will be used with a length $n = 7$ and $k = 4$ information symbols (the message). Thus there are $n - k = 3$ parity checks.

An (n,k)-code is generated with the help of k linearly independent words each with length n. We can represent this with the help of a generator matrix G, where the rows are formed by these k words. A channel code word x now arises according to

$$x = u \cdot G \tag{9.3}$$

where u is the message consisting of the information symbols $u_1, u_2, \ldots, u_k$ and G represents the $k \times n$ generator matrix. An example of a generator matrix G for a (7,4)-code is the following one:

$$G = \begin{bmatrix} 1000 & 011 \\ 0100 & 101 \\ 0010 & 110 \\ 0001 & 111 \end{bmatrix}.$$

A message $u = (0\ 1\ 1\ 1)$ yields the code word

$$x = ([0 \cdot 1 + 1 \cdot 0 + 1 \cdot 0 + 1 \cdot 0], [0 \cdot 0 + 1 \cdot 1 + 1 \cdot 0 + 1 \cdot 0], \ldots)$$

$$= (0,1,1,1,1,0,0) = (0\ 1\ 1\ 1\ 1\ 0\ 0).$$

Modulo two addition has been used here (0+0= 0, 0+1=1, 1+0=1, 1+1=0). In the field of coding, it is usual to denote the code words by a row vector and not as a column vector as is generally done in signal processing and linear algebra. This notation will be followed here as well.

The coding of a message can thus be considered as a multiplication of a $1 \times k$ vector (the message) and a $k \times n$ generator matrix G. This matrix must contain linearly independent rows, and for discrete memoryless channels, can always be written in a normal form according to

$$G = [I_k, A] \tag{9.4}$$

where I_k represents a $k \times k$ identity matrix. This has already been done for the previously given generator matrix for the (7,4)-code. A code where the generator matrix has this normal form is called a systematic code and has the property that the first k symbols of a code word are identical to the message. The remaining $n - k$ symbols are the parity checks c_i and follow from the submatrix A,

$$[x_1,\ldots,x_k] = [u_1,\ldots,u_k],$$

$$[x_{k+1},\ldots,x_n] = [u_1,\ldots u_k]\cdot A. \tag{9.5}$$

The generation of code words is therefore simpler with a systematic code than with a non-systematic code.

For the present generator matrix it yields for the parity check that

$$c_1 = u_2 + u_3 + u_4,$$

$$c_2 = u_1 + u_3 + u_4,$$

$$c_3 = u_1 + u_2 + u_4.$$

For code word x it follows thus that

$$x = (u_1, u_2, u_3, u_4, u_2 + u_3 + u_4, u_1 + u_3 + u_4, u_1 + u_2 + u_4).$$

Compare message $u = (0\ 1\ 1\ 1)$ and code word $x = (0\ 1\ 1\ 1\ 1\ 0\ 0)$.

The manner in which the parity checks are formed is in fact determined by the so-called *parity check matrix* H. For an (n,k) linear code C one finds an $(n-k) \times n$ matrix where each row specifies a parity symbol. For the previously given (7,4)-code we have

$$H = \begin{bmatrix} 0 & 1 & 1 & 1 & 1 & 0 & 0 \\ 1 & 0 & 1 & 1 & 0 & 1 & 0 \\ 1 & 1 & 0 & 1 & 0 & 0 & 1 \end{bmatrix}.$$

From this matrix the parity checks follow again: $c_1 = u_2 + u_3 + u_4$, $c_2 = u_1 + u_3 + u_4$ and $c_3 = u_1 + u_2 + u_4$.

Rewriting this yields

$$u_2 + u_3 + u_4 + c_1 = 0,$$

$$u_1 + u_3 + u_4 + c_2 = 0,$$

$$u_1 + u_2 + u_4 + c_3 = 0.$$

Obviously the parity check matrix is such that code word matrix multiplication with a code word x yields

$$x\,H^T = 0. \tag{9.6}$$

It seems that all code words generated by G meet the requirement $x \cdot \mathsf{H}^T = 0$. This gives a relationship between the matrices G and H. We call two vectors orthogonal if we find that their internal product is equal to zero.

For every $k \times n$ generator matrix G with k linearly independent rows one can find an $(n{-}k) \times n$ matrix H where the $(n - k)$ rows are linearly independent and each code word generated by G is orthogonal with respect to all the rows of H, or in other words $x \cdot h_i = 0$, with h_i a row of H. One sometimes also says that the subspace spanned by G is orthogonal with respect to the subspace spanned by H. For a systematic code with generator matrix $G = [I_k, A]$, the parity check matrix is given by

$$\mathsf{H} = [A^T, I_{n-k}].\tag{9.7}$$

By making use of the relation $x \cdot \mathsf{H}^T = 0$ it follows directly that the sum of two code words is also a code word.

$$(x + x') \cdot \mathsf{H}^T = x \cdot \mathsf{H}^T + x' \cdot \mathsf{H}^T$$

$$= 0 + 0 = 0.$$

9.3 Syndrome coding

Now consider a discrete memoryless channel. The assumption that the channel is memoryless means that we assume that errors can occur at arbitrary moments, without influencing each other. If a word y is received we regard this as the sum of a transmitted code word x and an error vector z. Thus we consider an additive channel (see Figure 9.2).

The error vector contains ones at the places where x and y differ, thus where an error has occurred during transmission. To decode we make use of the parity check matrix H, since $x \cdot \mathsf{H}^T = 0$ for a code word x. If errors have occurred then $z \neq 0$, so that it is reasonable then that $y \cdot \mathsf{H}^T \neq 0$. The vector $S = y \cdot \mathsf{H}^T$ is called the *syndrome*. The syndrome does not depend on x, but only on z since

$$S = y \cdot \mathsf{H}^T$$

$$= (x + z) \cdot \mathsf{H}^T$$

Figure 9.2. Additive channel.

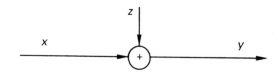

$$= x \cdot \mathsf{H}^T + z \cdot \mathsf{H}^T$$

$$= z \cdot \mathsf{H}^T. \tag{9.8}$$

Error detection can therefore simply take place by checking if $S = 0$. If $z = 0$ it is also the case that $S = 0$. It cannot be concluded with certainty that $z = 0$ from $S = 0$; certain error vectors let a code word go over into another code word which yields $S = 0$ as a result and thus yields an undetectable error.

If an error has been detected it must subsequently be determined where the error has occurred. We must therefore reconstruct the error pattern from S. We limit ourselves here to codes that are able to correct one error per code word. For the given example of a $(7,4)$-code there are then eight error patterns z possible, namely no errors ($z=0$) and seven times one error, (0000001) to (1000000).

The syndrome $S = y \cdot \mathsf{H}^T$ is a (1×3) vector with eight possible values: (000) to (111). We see that in this case a unique relation is possible between an error vector and the value of the syndrome. We can now make corrections after determining the syndrome S by finding the corresponding error vector in a table and adding it (modulo 2) to the received word y. The relationship between z and S follows by constantly adding one of the error patterns z to the code word (0000000) and calculating what S is. In this manner we find the table given below.

S	z
000	000 000 0
001	000 000 1
010	000 001 0
100	000 010 0
111	000 100 0
110	001 000 0
101	010 000 0
011	100 000 0

Example 9.2
Suppose that $y = (001\ 001\ 0)$ is received. Then the syndrome is

$$S = (0010010) \begin{bmatrix} 011 \\ 101 \\ 110 \\ 111 \\ 100 \\ 010 \\ 001 \end{bmatrix} = (100).$$

Hence, it follows that y is not a code word. From the table given above it follows that $S = (100)$ corresponds with an error vector $z = (0000100)$. By adding y and z the correct decoded code word follows:

$$x = (0010110). \qquad \qquad \triangle$$

In the example given here, the number of syndrome values is equal to the number of error vectors with a weight of at most one.

A more general approach makes use of the concept *coset*. For an (n,k)-code C and an arbitrary word a of length n, the set $a + C$ given by

$$a + C = \{a + x | x \in C\} \qquad (9.9)$$

is called a coset of C. All elements of a coset have the same syndrome because

$$(a + x) \cdot H^T = a \cdot H^T + x \cdot H^T = a \cdot H^T.$$

For a given syndrome $S = z \cdot H^T$ the possible solutions z lead to a coset of C. There exist 2^{n-k} of these cosets, each corresponding with one of the 2^{n-k} possible syndromes. Each coset consists of 2^k elements. There are thus in total $2^{n-k} \cdot 2^k = 2^n$ elements, which correspond with the 2^n possible received words y. Thus when the receiver has determined the syndrome, the search for the correct error vector (and thus the transmitted code word) has actually been brought back from 2^n to 2^k possibilities.

To determine the correct error vector from these 2^k possibilities we will determine the most probable one. For the binary symmetric memoryless channel with error probability p being considered by us here, the probability of a certain error vector is

$$p(z) = \prod_{i=1}^{n} p(z_i),$$

where

$$p(z_i = 0) = 1 - p,$$

$$p(z_i = 1) = p.$$

If an error vector contains t errors then the probability of such an error vector is

$$p(z) = p^t(1-p)^{n-t}.$$

Because it is usually true that $p < \frac{1}{2}$, $p(z)$ is a decreasing function of t so that the most probable error vector z is the vector with the smallest value of t, i.e. with the smallest weight. We thus come to the following decoding algorithm:

- Calculate the syndrome $S = y \cdot H^T$.
- Find the minimum weight vector z in the coset corresponding to S.
- Suppose $\hat{x} = y - z$.

It will become clear that the second step is the most complex. If k and $(n-k)$ are relatively small one can follow a so-called table search method. We will examine this method with the help of a (5,3)-code. The generator matrix for this code is

$$G = \begin{bmatrix} 1 & 0 & 0 & 1 & 1 \\ 0 & 1 & 0 & 0 & 1 \\ 0 & 0 & 1 & 0 & 1 \end{bmatrix},$$

whilst the parity check matrix is given by

$$H = \begin{bmatrix} 1 & 0 & 0 & 1 & 0 \\ 1 & 1 & 1 & 0 & 1 \end{bmatrix}.$$

There are four syndromes, namely 00, 01, 10, 11. The receiver can receive $2^5 = 32$ words, which we arrange in a 4×8 matrix. Each row contains the words that satisfy $S = z \cdot H^T$ for a certain syndrome. The matrix is given by the table below and is called the *standard matrix*.

S	coset leader							
00	00000	00101	01001	01100	10011	10110	11010	11111
01	00001	00100	01000	01101	10010	10111	11011	11110
10	00010	00111	01011	01110	10001	10100	11000	11101
11	10000	10101	11001	11100	00011	00110	01010	01111

The rows of this matrix are the cosets of C. The first row is the code itself. The vector with the smallest weight within the coset is located to the left and is called the coset leader $z(s)$. With the exception of the first row, a vector in the coset is equal to the coset leader plus the code word above the vector in question. Hence 01101 in the second row is the sum of 00001 and 01100. The second step in the decoding algorithm now becomes easy. The syndrome determines the coset. Since the coset leader $z(s)$ of the coset has the smallest weight and as such is most probable it is considered as the error pattern searched for. The supposed message then becomes $\hat{x} = y - z(s)$. In the case of received code word $y = (01101)$ this leads to $\hat{x} = (01100)$. Clearly, the error correction is only correct if the error pattern indeed was the coset leader.

In the other cases, which are less probable because of their weight, incorrect correction takes place. If one wants to be able to correct more error patterns, the number of syndrome values must become greater, which is only possible by increasing the number $n-k$ of parity checks.

The decoding method becomes too complex for large values of n and $n-k$. Other codes exist which allow relatively simple decoding methods for large lengths. An important class of codes makes use of a specification with the help of polynomials. This lies beyond the scope of this book and no further mention will be made of it here.

9.4 Hamming codes

Hamming codes form a class of codes which can correct one error. The simplest manner of depicting these codes is with the help of the parity check matrix. A Hamming code contains an $r \times (2^r - 1)$ parity check matrix and has all non-zero $(1 \times r)$ vectors as columns. As a consequence of this a Hamming code is a $(2^r - 1, 2^r - r - 1)$-code with length $n = 2^r - 1$, $k = 2^r - r - 1$ and r parity checks.

Hence for $r = 2$ we find

$$H_2 = \begin{bmatrix} 1 & 1 & 0 \\ 1 & 0 & 1 \end{bmatrix}.$$

The generator matrix for this code is

$$G_2 = [1\ 1\ 1],$$

which also indicates that we are also dealing with a repetition code. Hence, it can directly be seen that this code can correct one error.

For $r = 3$ we find in the same manner

$$H_3 = \begin{bmatrix} 0 & 1 & 1 & 1 & 1 & 0 & 0 \\ 1 & 0 & 1 & 1 & 0 & 1 & 0 \\ 1 & 1 & 0 & 1 & 0 & 0 & 1 \end{bmatrix},$$

where the columns are so arranged that a systematic code is obtained. The corresponding generator matrix is

$$G_3 = \begin{bmatrix} 1 & 0 & 0 & 0 & 0 & 1 & 1 \\ 0 & 1 & 0 & 0 & 1 & 0 & 1 \\ 0 & 0 & 1 & 0 & 1 & 1 & 0 \\ 0 & 0 & 0 & 1 & 1 & 1 & 1 \end{bmatrix}.$$

This (7,4)-Hamming-code has been used previously in this chapter as an example and can also correct one error. The decoding of Hamming codes is in particular a big advantage of these codes because it is rather simple. The coset leaders of a Hamming code are precisely the $n + 1 = 2^r$ vectors with weights ≤ 1, so that all error patterns with zero or one errors can be corrected.

By arranging the columns of the parity check matrix H in order of increasing binary value, the following simple decoding algorithm arises.

1. Determine the syndrome $S = y \cdot H^T$.

2a. If $S = 0$, then $\hat{x} = y$.

2b. If $S \neq 0$, then S gives the binary representation of the error position (under the assumption that exactly one error has been made).

3. $\hat{x} = y + z$.

Example 9.3
The matrix H_3 is arranged according to

$$H_3 = \begin{bmatrix} 0 & 0 & 0 & 1 & 1 & 1 & 1 \\ 0 & 1 & 1 & 0 & 0 & 1 & 1 \\ 1 & 0 & 1 & 0 & 1 & 0 & 1 \end{bmatrix}.$$

Suppose that $y = [1111011]$ is received. The syndrome now becomes $S = [101]$ and is therefore equal to the fifth column. An error has thus occurred at the fifth position. The correction now leads to the word $\hat{x} = [1111111]$.

9.5 Exercises

9.1 At a binary channel coding code words are contructed from two information symbols b_1 and b_2 and three parity check symbols (c_1, c_2 and c_3). The generator matrix is as follows:

$$G = \begin{bmatrix} 1 & 0 & 1 & 1 & 1 \\ 0 & 1 & 1 & 0 & 1 \end{bmatrix}.$$

a. Determine the code words of the code, their weights and their Hamming distances.
b. Determine the parity check matrix H.
c. Determine the syndrome for the following error patterns:

(01000), (00101), (10010), and (11111).

d. A code word (11010) is generated, which is distorted with the following error pattern: (10010). Determine the received code word and which correction will be applied. Explain the decision.

9.2. A binary code is constructed by adding three parity check symbols to three information symbols (denoted by b_1, b_2 and b_3). For these code symbols

$$x_1 = b_1,$$
$$x_2 = b_2,$$
$$x_3 = b_3,$$
$$x_4 = b_2 + b_3,$$
$$x_5 = b_1 + b_3,$$
$$x_6 = b_1 + b_2.$$

a. Determine the code words of this code.
b. Determine the control or parity check matrix H.
c. Determine the generator matrix G.
d. Determine the Hamming distance of the code.
e. How many errors can this code detect and how many errors can it correct?

9.3. Consider a binary linear block code where for every three information symbols one parity check symbol is added so that the amount of '1' is even in every code word.
a. Give the control matrix H of the code.
b. Give the generator matrix G of the code.
c. Determine the Hamming distance of the code.

d. How many errors can be detected and how many errors can be corrected? Explain your answer.

e. Give the standard matrix in relation to the cosets that play a role in the decoding with the aid of a syndrome.

9.4. A binary linear block code has the following parity check matrix:

$$H = \begin{bmatrix} 1 & 1 & 1 & 0 & 1 & 0 & 0 \\ 1 & 1 & 0 & 1 & 0 & 1 & 0 \\ 1 & 0 & 1 & 1 & 0 & 0 & 1 \end{bmatrix}.$$

a. How many code words does this code contain?

b. Give a systematic generator matrix of the code.

c. After transmitting two code words with a binary symmetric channel (with error probability < 0.5) the vectors $y_1 = (1000001)$ and $y_2 = (1001100)$ are received.

– Calculate the syndrome for both vectors.

– Determine for both vectors, with the aid of syndrome decoding, the most probable code word transmitted.

– What are the most probable information symbols?

9.6 Solutions

9.1. *a.* Applying $x = u \cdot G$, with u a message consisting of b_1 and b_2, leads to the following code words.

x	$w(x)$
10111	4
01101	3
11010	3
00000	0

The Hamming distances between the code words are presented in the following symmetric matrix.

$$\begin{bmatrix} 0 & 3 & 3 & 4 \\ 3 & 0 & 4 & 3 \\ 3 & 4 & 0 & 3 \\ 4 & 3 & 3 & 0 \end{bmatrix}.$$

b. The parity check matrix can be determined by $x \cdot H^T = 0$ for each x. The general form of H is

$$H = \begin{bmatrix} a & b & 1 & 0 & 0 \\ c & d & 0 & 1 & 0 \\ e & f & 0 & 0 & 1 \end{bmatrix}.$$

With the help of four code words and $x \cdot H^T = 0$ we obtain

$$\begin{aligned}
\text{code word 1:} \quad & a = 1 \\
& c = 1 \\
& e = 1
\end{aligned}$$

$$\begin{aligned}
\text{code word 2:} \quad & b = 1 \\
& d = 0 \\
& f = 1
\end{aligned}$$

$$\begin{aligned}
\text{code word 3:} \quad & a + b = 1 \\
& c + d = 1 \\
& e + f = 0
\end{aligned}$$

$$\text{code word 4:} \quad \quad —$$

The resulting matrix H becomes

$$H = \begin{bmatrix} 1 & 1 & 1 & 0 & 0 \\ 1 & 0 & 0 & 1 & 0 \\ 1 & 1 & 0 & 0 & 1 \end{bmatrix}.$$

c. If z is an error vector, the syndrome S equals $S = z \cdot H^T$. For the four error vectors mentioned we obtain $S = [101]$.

d. The transmitted code word x is (11010), the error vector is (10010), thus the received code word is $y = x + z = (01000)$ with syndrome as given in c.

From the vectors given in c, (01000) has the smallest weight, this vector is then considered as the error vector. Hence, the supposed code word becomes $\hat{x} = y - z(s) = (00000)$, which is incorrect in this case.

9.2. *a.* There are 8 messages. Their corresponding code words are given as follows:

message	code words
000	000 000
001	001 110
010	010 101
011	011 011
100	100 011
101	101 101
110	110 110
111	111 000

b. The control matirx H should satisfy $x \cdot H^T = 0$, for each code word x. It follows that

$$H = \begin{bmatrix} 0 & 1 & 1 & 1 & 0 & 0 \\ 1 & 0 & 1 & 0 & 1 & 0 \\ 1 & 1 & 0 & 0 & 0 & 1 \end{bmatrix}.$$

c. The generator matrix G can be derived from $G \cdot H^T = 0$. The result is

$$G = \begin{bmatrix} 1 & 0 & 0 & 0 & 1 & 1 \\ 0 & 1 & 0 & 1 & 0 & 1 \\ 0 & 0 & 1 & 1 & 1 & 0 \end{bmatrix}.$$

This result can also be obtained by writing H as $H = [A^T, I_{n-k}] = [A^T, I_3]$ and noting that then $G = [I_3, A]$. Another possibility is deducing G from $x = u \cdot G$.

d. The smallest Hamming distance between the code words is 3. Thus the Hamming distance d of the code is equal to 3.

e. For Hamming distance $d = 3$, $d - 1 = 2$ errors can be detected and $t = \lceil (d - 1)/2 \rceil = 1$ error can be corrected.

9.3. *a.* Clearly, for the parity check symbol $c_1 = u_1 + u_2 + u_3$ holds. Thus, the control matrix is given by

$$H = [1 \ 1 \ 1 \ 1].$$

b. If $H = [A^T, I_{n-K}]$ then $G = [I_k, A]$. Here, the generator matrix is the following.

$$G = \begin{bmatrix} 1 & 0 & 0 & 1 \\ 0 & 1 & 0 & 1 \\ 0 & 0 & 1 & 1 \end{bmatrix}.$$

c. The code words are given in the following table.

message u	code word x
000	0000
001	0011
010	0101
011	0110
100	1001
101	1010
110	1100
111	1111

From this a Hamming distance $d = 2$ follows.

d. Since $d = 2$, 1 ($= d - 1$) error can be detected. No errors can be corrected, since $t = \lceil (d - 1)/2 \rceil = 0$.

e. It is the case that $S = y \cdot H^T$. There are two syndromes: $S = 0$ or $S = 1$. We find

S	coset leader							
0	0000	0011	0110	1100	0101	1010	1001	1111
1	0001	0010	0111	1101	0100	1011	1000	1110

9.4. *a.* It may be mentioned here that considering the matrix H it may be concluded that we are dealing with a Hamming code. The parity check matrix is an $(n - k) \times n$ matrix, thus $n = 7$ and $n - k = 7 - k = 3$, which leads to $k = 4$. There are now $2^k = 16$ messages.

b. The generator matrix G derived from H is

$$G = \begin{bmatrix} 1 & 0 & 0 & 0 & 1 & 1 & 1 \\ 0 & 1 & 0 & 0 & 1 & 1 & 0 \\ 0 & 0 & 1 & 0 & 1 & 0 & 1 \\ 0 & 0 & 0 & 1 & 0 & 1 & 1 \end{bmatrix}.$$

c. For code word y_1 the syndrome is equal to $S_1 = y_1 \cdot H^T = (110)$. This correponds to the second column of the parity check matrix H. Thus, probably the second bit of the received code word should be corrected.

Therefore, the assumed code word is (1100001) and the transmitted message (1100).

Code word y_2 has a syndrome $S_2 = y_2 \cdot H^T = (000)$. Due to this syndrome value, it is assumed that there were no errors and that the received code word is correct. The transmitted message is (1001).

10

Cryptology

10.1 Cryptography and cryptanalysis

Another scientific area wherein the results from information theory are applied is *cryptology*. Since the term cryptology is a contraction of the Greek words 'kruptos' and 'logos', it literally means the study of hiding. As such it deals among others with the development of methods for the *encipherment* and *decipherment* of messages.

The desire to bring messages in such a form that the original message cannot be understood by those for whom the message is not intended is very old. Use has been made of secret codes since the beginning of recorded history. From early history up until about World War Two secret codes were mainly used in military and diplomatic environments. In recent years developments have taken place which have caused the demand for methods of enciphering messages outside of military or diplomatic environments to also grow.

To give an example: consider cable television, where in principle only paying customers should be able to receive the programmes transmitted by the television station, whilst reception for those who have not paid must be impossible. This can be achieved if the images are transmitted in an enciphered form, while a system is placed at the receiver which deciphers the image again. An important field which cannot exist without cryptology is electronic banking. Magnetic cards, bank passes etc. for monetary transactions, they all make use of cryptographic tools.

Think of problems concerning the protection of privacy. Because of the expansion of automatisation there are a continually increasing number of systems containing data bases with personal information, e.g. medical data bases, juridical data bases etc. In many of these cases it is desirable to protect the stored data against unwanted consultation. One may also want to prevent eavesdropping during the transportation of data (text, speech, video) via telecommunication networks, etc.

Within cryptology, distinction is made between two disciplines, namely those of *cryptography* and *cryptanalysis*. Cryptography is the area of cryptology that is concerned with the development and the study of methods and methodologies of enciphering. Here usually one makes use of secret keys. Only those who have the secret key can decipher the enciphered information. For any other this is impossible or near impossible. Cryptanalysis is the area of cryptology that seeks to develop techniques in order to decipher enciphered messages; that means without any a priori knowledge with respect to e.g. the key ("hacking").

Obviously, it is not possible to go into the many facets of cryptography and cryptanalysis within the scope of this book. In this chapter we will therefore limit ourselves to a general introduction to the field and pay some attention to those aspects of cryptography and cryptanalysis where the emphasis lies on the application of concepts found in information theory.

10.2 The general scheme of cipher systems

In Figure 10.1 a general outline of a *cipher system* is given. A source which generates messages *M*, denoted by *plaintext*, is found on the side of the transmitter. The plaintext is transformed into a *ciphertext*, which is denoted by the letter *C*, through the use of some enciphering method. The enciphering operation can be regarded as a transformation *T*, which transforms *M* into *C*. We will give a number of simple examples of enciphering methods in one of the subsequent sections.

There are a number of possibilities for the transformation, depending on the chosen key *K*. This key originates from a set of possible keys. Thus we have

$$C = T_K(M).$$

Figure 10.1. Outline of a cipher system.

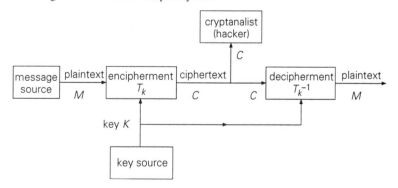

Deciphering takes place on the receiving side by processing the received ciphertext with the inverse transformation T_{kK}^{-1}, that is $M = T_K^{-1}(C)$. It is usually assumed that the transformation T itself is known, but not the key. The cryptanalyst has the task of either discovering the key from the ciphertext or discovering the plaintext directly. In future we will assume that the key is different for each new message. This means that the plaintext, the ciphertext and the key can all be regarded as stochastic quantities. It is here that a link can be made to information theory as we will see.

Three types of cryptanalytical attacks on cryptosystems can be distinguished, corresponding with the nature of the information that the cryptanalyst has access to. These three sorts of attacks are
– ciphertext-only attack,
– known-plaintext attack,
– chosen-plaintext attack.

In the event that the cryptanalyst only has the ciphertext (the enciphered message), he must try to decipher the underlying message (plaintext) by analysing the structure and possible statistical features present within the ciphertext or, more importantly, he must try to find the key. This is the ciphertext-only attack.

The situation where besides having the ciphertext, the cryptanalyst also has some information about the corresponding plaintext is more favourable than the previous situation. Through the knowledge of combinations of pieces of the ciphertext and plaintext the cryptanalyst can now try to decipher the part of the ciphertext for which the corresponding plaintext is not known (known-plaintext attack). Situations where the cryptanalyst has the ciphertext as well as a part of the corresponding plaintext can arise when the cryptanalyst has succeeded in penetrating into the cryptosystem or to the user of the system. For automatic monetary traffic for example, information must be present regarding the person transferring the money as well as the bank where the money must be transferred to with every transaction. If through inside information the cryptanalyst knows where information concerning the bank, account number etc. is hidden in the ciphertext, he can try to decipher the rest of the ciphertext on the basis of this knowledge.

The most favourable situation for the cryptanalyst arises when the cryptanalyst can choose the plaintext himself and can compare this with the resulting ciphertext (chosen-plaintext attack). This situation can arise in the case of cryptanalysis of ciphersystems for word processors among others.

When applying cryptographic methods one naturally gives the preference to developing a system that is proof against all three types of attack. In practice this proves to be difficult to realise. A system that appears to be safe against a ciphertext-only attack need not be so against an attack by means of chosen plaintext. In practice, a system that can survive an attack on the basis of a chosen plaintext is more highly regarded than a system that can only survive an attack on the basis of ciphertext only.

A number of applications where cryptography is applied have already come up in previous sections. Generally speaking, one can divide the applications up into two groups, namely applications regarding storage and applications concerning transmission.

In the case of storage, one should think of the storage of data in computer systems: on disk or on magnetic tape. The method through which the data or software is stored is often known in this case, but not the key. A cryptanalytic attack is attractive here since data is often stored for long periods of time. The cryptanalyst thus has ample time to find the key.

In the case of transmission (telephone, TV, via cable or satellite connections) the enciphered message is usually only available to the cryptanalyst for a very short period of time, as opposed to storage, and also only at the moment of transmission. A change of key can moreover more easily take place than in the case of storage. Furthermore, with telecommunication the messages are often only of value for a limited amount of time, since the contents may be out of date after a certain amount of time for example (think of news, weather information etc.). Cryptographic methods which may not be absolutely safe may still be attractive provided that the value of the message has greatly decreased within the amount of time that is necessary for an eavesdropper to decipher it.

10.3 Cipher systems

Two basic enciphering methods can be distinguished:
- *stream enciphering*,
- *block enciphering*.

With stream enciphering the message is regarded as being built up from a successive sequence of a number of loose elements. The elements can be thought of as being letters but also binary or ASCII characters for example. Characteristic of stream enciphering is that the text is enciphered element by element. Use is often made of shift-registers with this method of enciphering.

In the case of block enciphering a number of elements are continually taken together and enciphered as one whole. An example of block enciphering is the DES-algorithm developed by IBM in 1968-1975 (DES = Data Encryption Standard), which is at present one of the most widely used algorithms for block enciphering. DES assumes that the data is delivered in binary form. It is designed for the enciphering and deciphering of blocks consisting of 64 bits. The utilised key amounts to 64 bits, of which only 56 are actually used; the remaining 8 bits are parity checks. The number of keys thus comes to about $7.2 \cdot 10^{16}$.

The two basis systems for block enciphering are the *transposition* and the *substitution cipher systems*. These methods of enciphering have had a very long history. Most of these methods have lost much of their meaning over the years. They were often used up until the Second World War, but are applied less and less often since the cryptanalyst gained access to computers. That does not mean that the discussion of classic cipher systems is only still important in a historical sense. On the contrary, the classic systems may not be used very often by themselves, but they are still used within the more modern cryptosystems, as building stones. A modern cryptographic algorithm such as DES can be regarded as a series of transpositions and substitutions, for example.

A transposition cipher is characterised by the fact that the symbols in which the plaintext is rendered are not altered, but that only their sequence is altered. In the case of substitution ciphers the sequence is not altered but the symbols are; the original symbols in the plaintext are replaced by other symbols.

Transposition ciphers
With transposition ciphers only the ordering of the symbols or elements of which the plaintext is constructed is altered. This is done on a block basis. Consider the following example.

plaintext   : the invasion will begin
division into blocks : thein vasio nwill begin
ciphertext   : ehnti saovi iwlnl genbi

Although nowadays transposition and substitution are mainly performed on bit streams, our examples in this chapter are related to language because of clearness. The plaintext in the example is partitioned into blocks consisting of 5 letters. One then speaks of a *period length* equal to 5 in this case. The letters are rearranged within each block according to the key 3 2 5 1 4. Thus with respect to the original block, the third and second letters are placed in the first and second positions, the fifth and first letters are placed in the third and fourth positions respectively, while the fourth letter is placed in position five.

The encipherment of messages according to the transposition cipher can in fact be thought of as the application of column transposition, as will become apparent from the following example. Instead of placing blocks of 5 letters beside each other, they are now placed beneath each other.

plaintext: the invasion will begin

key word: key 32514

```
thein        ehnti
vasio        saovi
nwill        iwlnl
begin        genbi
```

Ciphertext: ehnti saovi iwlnl genbi

It is clear that the cryptogram can be obtained by exchanging the columns in accordance with the keyword.

We can now describe the transposition cipher in more general terms as follows. For a period length of T the total number of keys is equal to $T!$, or actually $T!-1$, since there is one key that will yield a ciphertext that is identical to the plaintext.

In the example given in this section the period length is equal to 5, which means that there are $5!-1 = 119$ possible keys. This is in practice useless. If the cryptanalyst knows the period length, he will quickly be able to decipher the cryptogram.

In general the cryptanalyst faces two problems with transposition ciphers with large period. First of all he must try to find the period length. To do this he should in effect try all the numbers n and T that satisfy $L = nT$, where L is the message length. If the possibility that dummy letters have been added to the plaintext is not to be excluded, one must consider still more combinations of n and T.

The second problem, once the period length is known, is to find the key in a structured manner, thus without having to try out all of the possible permutations.

In the case of language plaintexts the cryptanalyst can make use of language characteristics to overcome these two problems. If the letter frequencies of a language are considered, we find that some letters appear more often than other letters (see Figure 10.2).

The same is true if one does not look at the relative frequencies of the individual letters, but at the frequencies of pairs of letters (bigrams). It turns out that when considering text, vowels are often surrounded by consonants and vice versa. This means that vowels will be fairly evenly spread out over the text. This means that when determining the period length one should look for that T for which the resulting columns show the most regular distribution of vowels over the columns. During the decipherment one then subsequently places next to each other those columns which will yield a large number of frequently occurring letter pairs. Clearly, this is only possible if the underlying plaintext is language. Otherwise, it is more difficult.

Figure 10.2. Letter frequency quotients for various different languages.

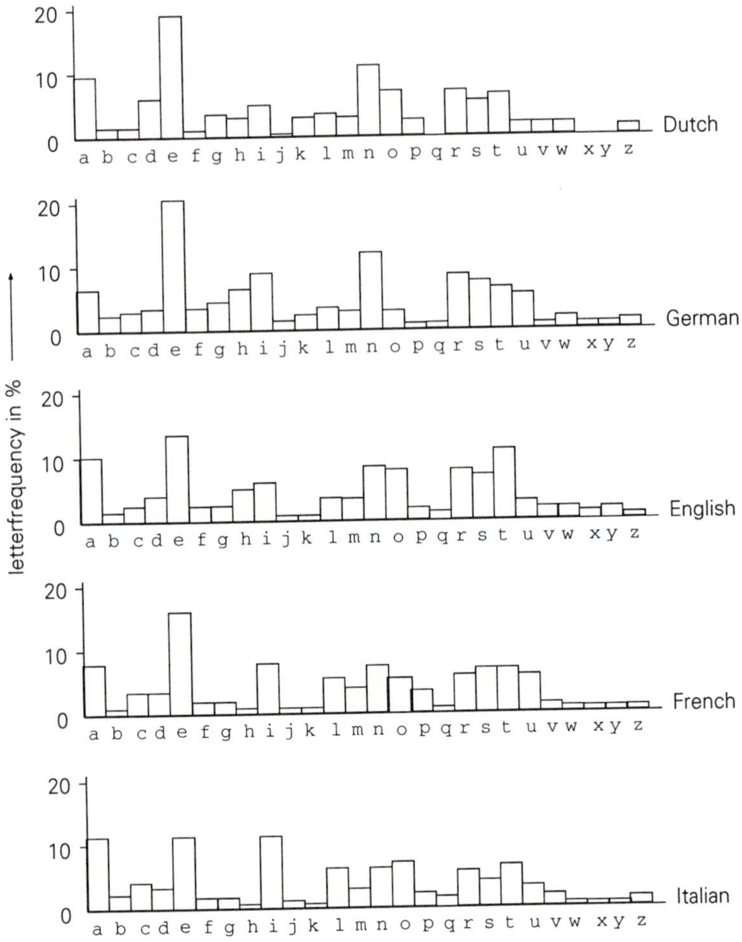

Substitution ciphers

In substitution ciphers the symbols of the plaintext are replaced by other symbols. Suppose that one has an alphabet consisting of 26 letters, then a substitution cipher system can in general terms be described as follows.

alphabet with regard to plaintext: $A = [a_1,...,a_{26}]$
alphabet with regard to ciphertext: $B = [b_1,...,b_{26}]$
plaintext: $a_3, a_{23}, a_9, a_{17}, a_4$
ciphertext: $b_3, b_{23}, b_9, b_{17}, b_4$.

The simplest substitution is the *Caesar* substitution, which takes its name from Julius Caesar. The substitution alphabet consists of the displaced original alphabet. A shift of 3 places has been applied in the following example.
Original alphabet A:

    a b c d e f g h i j k l m n o p q r s t u v w x y z

Substitution alphabet B:

    d e f g h i j k l m n o p q r s t u v w x y z a b c

plaintext   : the invasion will begin
ciphertext : wkh lqydvlrq zloo ehjlq

Characteristic of the Caesar substitution is that the alphabet remains the same. The number of keys amounts to only 26, so that the ciphertext can easily be cracked. It is only necessary to know the corresponding letter in the ciphertext for one letter, and the system is cracked. Such a letter can easily be found if the message is large enough. One just has to look for the most frequently occurring letter in the cryptogram; there is a large probability that this letter corresponds with the letter e in the plaintext.

If one now takes an alphabet where the letters are placed in an arbitrary order instead of an alphabet with displaced letters, then the number of keys becomes 26! (26 factorial). This makes deciphering considerably more difficult than in the case of Caesar substitution. An example is:
Original alphabet A:

    a b c d e f g h i j k l m n o p q r s t u v w x y z

Substitution alphabet B:

    e s t v f u z g y x b h k w c i r j a l m p d q o n

plaintext   : the invasion will begin
ciphertext : lgf ywpeaycw dyhh sfzyw

Despite the 26! possible keys, such substitutions are still relatively easy to solve. As we know, language is redundant to a high degree. Furthermore, the most frequently appearing letters such as e, t, a, n, etc. can always be easily found in the ciphertext on the basis of the frequency distribution of the letters in the ciphertext.

One can conclude that the substitution methods, as described above, are not that powerful. This is because the language's features are still relatively easy to extract from the ciphertext. Therefore often more than one substitution is applied. One speaks of *polyalphabetic substitution* in this case. A well known example of a polyalphabetic substitution is the Vigenère system. What this comes down to is that not one, but a number of Caesar substitutions are applied. The first letter in the plaintext is shifted over 20 places for example, the second over 17 places etc.

One often makes use of the so-called Vigenère-tableau (see Figure 10.3) and a chosen keyword. In the Vigenère-tableau, the letters of the alphabet of the plaintext are located in the top row. The first column contains the possible letters of the keyword. Encipherment now proceeds as follows.

Figure 10.3. The Vigenère-tableau.

```
A  B  C  D  E  F  G  H  I  J  K  L  M  N  O  P  Q  R  S  T  U  V  W  X  Y  Z
B  C  D  E  F  G  H  I  J  K  L  M  N  O  P  Q  R  S  T  U  V  W  X  Y  Z  A
C  D  E  F  G  H  I  J  K  L  M  N  O  P  Q  R  S  T  U  V  W  X  Y  Z  A  B
D  E  F  G  H  I  J  K  L  M  N  O  P  Q  R  S  T  U  V  W  X  Y  Z  A  B  C
E  F  G  H  I  J  K  L  M  N  O  P  Q  R  S  T  U  V  W  X  Y  Z  A  B  C  D
F  G  H  I  J  K  L  M  N  O  P  Q  R  S  T  U  V  W  X  Y  Z  A  B  C  D  E
G  H  I  J  K  L  M  N  O  P  Q  R  S  T  U  V  W  X  Y  Z  A  B  C  D  E  F
H  I  J  K  L  M  N  O  P  Q  R  S  T  U  V  W  X  Y  Z  A  B  C  D  E  F  G
I  J  K  L  M  N  O  P  Q  R  S  T  U  V  W  X  Y  Z  A  B  C  D  E  F  G  H
J  K  L  M  N  O  P  Q  R  S  T  U  V  W  X  Y  Z  A  B  C  D  E  F  G  H  I
K  L  M  N  O  P  Q  R  S  T  U  V  W  X  Y  Z  A  B  C  D  E  F  G  H  I  J
L  M  N  O  P  Q  R  S  T  U  V  W  X  Y  Z  A  B  C  D  E  F  G  H  I  J  K
M  N  O  P  Q  R  S  T  U  V  W  X  Y  Z  A  B  C  D  E  F  G  H  I  J  K  L
N  O  P  Q  R  S  T  U  V  W  X  Y  Z  A  B  C  D  E  F  G  H  I  J  K  L  M
O  P  Q  R  S  T  U  V  W  X  Y  Z  A  B  C  D  E  F  G  H  I  J  K  L  M  N
P  Q  R  S  T  U  V  W  X  Y  Z  A  B  C  D  E  F  G  H  I  J  K  L  M  N  O
Q  R  S  T  U  V  W  X  Y  Z  A  B  C  D  E  F  G  H  I  J  K  L  M  N  O  P
R  S  T  U  V  W  X  Y  Z  A  B  C  D  E  F  G  H  I  J  K  L  M  N  O  P  Q
S  T  U  V  W  X  Y  Z  A  B  C  D  E  F  G  H  I  J  K  L  M  N  O  P  Q  R
T  U  V  W  X  Y  Z  A  B  C  D  E  F  G  H  I  J  K  L  M  N  O  P  Q  R  S
U  V  W  X  Y  Z  A  B  C  D  E  F  G  H  I  J  K  L  M  N  O  P  Q  R  S  T
V  W  X  Y  Z  A  B  C  D  E  F  G  H  I  J  K  L  M  N  O  P  Q  R  S  T  U
W  X  Y  Z  A  B  C  D  E  F  G  H  I  J  K  L  M  N  O  P  Q  R  S  T  U  V
X  Y  Z  A  B  C  D  E  F  G  H  I  J  K  L  M  N  O  P  Q  R  S  T  U  V  W
Y  Z  A  B  C  D  E  F  G  H  I  J  K  L  M  N  O  P  Q  R  S  T  U  V  W  X
Z  A  B  C  D  E  F  G  H  I  J  K  L  M  N  O  P  Q  R  S  T  U  V  W  X  Y
```

The keyword is placed under the plaintext as indicated in the example below. The letters of the cryptogram are now found by taking the letter in the tableau that lies on the intersection of the column with the letter from the plaintext and the row with the letter from the keyword.

Plaintext : the invasion will begin

key : *rad ioradior adio radio*

ciphertext : khh qbmavqce wltz sejqb

The conclusion is that a letter in the plaintext can be represented by different letters in the ciphertext, depending on the letters in the keyword. Because of this, language characteristics are better concealed than in the previously mentioned methods.

The number of monoalphabetic substitutions, which form the basis of the Vigenère system, is equal to the length of the keyword. For the example given above the number of monoalphabetic substitutions used amounts to 5. This means that 5 rows have been used from the tableau.

Knowledge of the length of the keyword will clearly be of great help to the cryptanalyst in solving the cryptogram.

The user of the system will generally strive to use as many of the tableau's rows as possible during encipherment. A method of achieving this is to make use of the plaintext itself as well as the keyword, as is illustrated in the following example.

plaintext : the invasion will begin

key : *rad iotheinv asio nwill*

ciphertext : khh qbohwqbi watz oaoty

After using the key "radio" the plaintext itself is used as key.

One final remark. The problem with the application of transposition and substitution ciphers is to conceal the characteristics of underlying plaintext as well as possible. One of the solutions is to make certain that the letters or symbols in the ciphertext exhibit a uniform distribution. This can be achieved by, before applying a transposition or substitution method, enciphering the letters in the plaintext using a method such as Huffman, as has already been treated in the previous chapters. An optimal source coding will have as result that all the code symbols have the same probability of occurring.

10.4 Amount of information and security

It is of great importance for the application of cryptographic systems to have an impression of the security of the system used. By making use of concepts within information theory an attempt will be made to obtain insight into the question of what a safe cryptosystem actually is.

By regarding the messages in plaintext as elements of a set of possible messages generated by the source, each with its own probability of occurring, it is possible to speak of the amount of information in the plaintext.

The amount of information in the plaintext is formulated as follows:

$$H(M) = -\sum_{i=1}^{n} p(M_i) \log p(M_i), \tag{10.1}$$

where $p(M_i)$, $i=1,...,n$, are the probabilities of occurrence of the plaintext messages M_i. We can speak of the amount of information of the ciphertext, denoted by $H(C)$, and the amount of information $H(K)$ with regard to the keys, in the same manner.

In an analogous manner, we can also speak of the conditional amount of information. $H(K/C)$ is then the amount of information or uncertainty with regard to the keys if we have knowledge about the ciphertext C, also known as the *key equivocation*. This can be defined as follows. Let K_h, $h = 1,...,l$, be the set of keys and C_j, $j = 1,...,m$, the possible cryptograms, then

$$H(K/C) = -\sum_{h=1}^{l} \sum_{j=1}^{m} p(K_h, C_j) \log p(K_h/C_j). \tag{10.2}$$

Likewise, H(M/C) is the amount of information or uncertainty concerning the plaintext M for a given ciphertext, also called the *message equivocation*. $H(M/C,K)$ can similarly be regarded as the amount of information with regard to the plaintext if both the ciphertext and the key are known.

Since the plaintext is unambiguously determined by the ciphertext and the key, we have

$$H(M/C,K) = 0. \tag{10.3}$$

If we have access to the ciphertext and the key, then it is also possible to find the plaintext: the uncertainty about M is then equal to 0.

$H(K/M,C)$, the so-called *key appearance equivocation*, is the amount of information with regard to the key given the plaintext and ciphertext.

Theorem 10.1
The following equality holds for the key appearance equivocation:

$$H(K/M,C) = H(K/C) - H(M/C).$$ (10.4)

Proof
On the basis of previously given relations for the amount of joint information we find that the amount of information in the plaintext, ciphertext and key is equal to

$$H(M,C,K) = H(M/C,K) + H(C,K)$$

$$= H(K/M,C) + H(M,C).$$ (10.5)

Bearing in mind that it is also the case that

$$H(C,K) = H(K/C) + H(C),$$

and

$$H(M,C) = H(M/C) + H(C),$$

it then follows from equation (10.5) that

$$H(M/C,K) + H(K/C) = H(K/M,C) + H(M/C).$$ (10.6)

We have already seen, above, that $H(M/C,K) = 0$ (see equation (10.3)). Equation (10.4) then follows with the help of equation (10.6) .   □

Theorem 10.1 gives rise to some interesting interpretations. From the user's point of view one strives for a large value of $H(K/M,C)$. If the cryptanalyst has at his disposal both the plaintext and the ciphertext, then one should in any case have made certain that the uncertainty about which key has been used is as large as possible. It can be concluded on the basis of equation (10.4) that a large value of $H(K/M,C)$ can be achieved by making sure that $H(M/C)$ assumes a small value. A small value of $H(M/C)$, however, means that there is little uncertainty about the plaintext M if one only has access to the ciphertext C. We rather want to avoid this, however. We have in fact run up against a dilemma because of this. A large uncertainty with regard to the key goes at a cost to the uncertainty about the transmitted plaintext. Conversely, a large uncertainty with regard to the plaintext will be accompanied by a small uncertainty with regard to the key.

With the help of this information theoretic consideration other important conclusions can be reached. Let $I(M;C)$ be the mutual information between plaintext and ciphertext, which is defined by

$$I(M;C) = H(M) - H(M/C)$$

$$= H(C) - H(C/M). \tag{10.7}$$

From the user's point of view of the cryptosystem, one will strive to make $I(M;C)$ as small as possible, since the mutual information is a measure of the mutual independence and this must be small in this case. If the ciphertext gives absolutely no information about the plaintext, then $H(M/C) = H(M)$. It then follows from equation (10.7) that the mutual information between the plaintext and the ciphertext will then become equal to 0.

We speak of an *absolutely secure cryptosystem* if

$$I(M;C) = 0. \tag{10.8}$$

An important relation for encipherment systems is given in the following theorem.

Theorem 10.2

Let $I(M;C)$ be the mutual information between plaintext and ciphertext, then

$$I(M;C) \geq H(M) - H(K). \tag{10.9}$$

Proof

To prove this relation we reconsider equation (10.4). Since $H(K/M,C) \geq 0$, it can be derived from equation (10.4) that

$$H(K/C) \geq H(M/C). \tag{10.10}$$

As by definition

$$H(K) \geq H(K/C),$$

it follows from formula (10.10) that

$$H(K) \geq H(M/C). \tag{10.11}$$

If we combine this inequality with equation (10.7) then formula (10.9) immediately follows. $\qquad\qquad\square$

The relationship in formula (10.9) really says that a set of keys that contain little information (on average) makes a large mutual information possible between plaintext and ciphertext.

Absolute security, that is $I(M;C) = 0$, can only be achieved in the case where

$$H(K) \geq H(M). \tag{10.12}$$

The information in the key must therefore be at least as large as the information in the plaintext.

10.5 The unicity distance

In the previous section, the length of the ciphertext intercepted by the cryptanalyst was not taken into account. Yet this is of substantial importance. Suppose that the plaintext is in language. As has previously been mentioned language shows certain statistical characteristics. On the basis of these underlying statistical characteristics it is often possible for the cryptanalyst to decipher parts or all of the ciphertext.

It is generally true that the longer the ciphertext in the possession of the cryptanalyst, the greater the probability that he will be able to find the key. If we denote a ciphertext of length L as C^L, then the key equivocation $H(K/C^L)$ will decrease for increasing values of L, because if L increases the uncertainty about the key used decreases. This is illustrated in Figure 10.4. The key equivocation will even become 0 at a given moment, when the key can be determined from the ciphertext with certainty. The maximum value of the key equivocation will be equal to $H(K)$. This value will occur when the probabilities of a key given a ciphertext are equal to the probability of that key itself. We then have maximum uncertainty with regard to the utilised key.

A similar consideration applies to the message equivocation $H(M^L/C^L)$. There is one difference, however: the total number of messages will still be

Figure 10.4. Key, message and key appearance equivocation as functions of L.

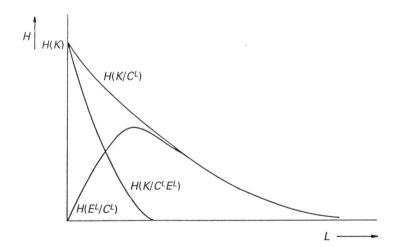

small for small values of L, which has as a consequence that the message equivocation will then also assume a small value. If L increases, however, the number of possible messages will quickly increase and therefore also the message equivocation. If L assumes a large enough value then the ciphertext or cryptogram will eventually contain enough information to stop the number of most probable messages from increasing further. Shortly after this point the message equivocation will even coincide with the key equivocation, because the cryptogram will then contain all the information to determine the key from the found plaintext with the same certainty and vice versa.

The course of the key appearance equivocation $H(K/C^L,M^L)$ is also depicted in the same figure. The key appearance equivocation will clearly approach zero faster than the key equivocation, because in the case of the key appearance equivocation it is assumed that the cryptanalyst also has the plaintext as well as the ciphertext. This extra knowledge will allow the cryptanalyst to determine the key more quickly on average.

The key appearance equivocation is a measure for the strength of a cipher system under a known-plaintext attack for the key, while the key and message equivocation are a measure for the strength of a cipher system under a ciphertext-only attack for the key and message respectively.

From the above it has become clear that as the length of the received ciphertext becomes larger, the probability that the cryptanalyst will be able to find the key or plaintext will increase.

Theorem 10.3

Let ε be the number of different symbols in a message or ciphertext of length L. For the key equivocation we now have

$$H(K/C^L) \geq H(K) - D_L, \tag{10.13}$$

where D_L is called the absolute redundancy which is defined by

$$D_L = L \log(\varepsilon) - H(M^L). \tag{10.14}$$

Proof

Because an unambiguous relation exists between plaintext and ciphertext it will always be the case that

$$H(K,C^L) = H(K,M^L).$$

Under the assumption that the key is independent of the message source, we therefore have for the key equivocation

$$H(K/C^L) = H(K,C^L) - H(C^L)$$
$$= H(K,M^L) - H(C^L)$$
$$= H(K) + H(M^L) - H(C^L). \qquad (10.15)$$

As ε is the number of different symbols in a message or ciphertext, there are ε^L possible messages or cryptograms of length L in total. On the basis of the properties of the information measure it can be proven that

$$H(C^L) \leq L \log(\varepsilon).$$

With this and with the help of equation (10.15) it can be found that

$$H(K/C^L) \geq H(K) + H(M^L) - L \log(\varepsilon). \qquad (10.16)$$

The term $[H(M^L) - L \log(\varepsilon)]$ in equation (10.16) is minus the redundancy D_L, so that equation (10.16) goes over into equation (10.13). □

The absolute redundancy can be regarded as a measure for the extent to which the actual message source differs from a source where each message has an equal probability of occurring. It is closely related to the definition of redundancy in equation (2.3), which in fact is a measure of relative redundancy.

An interpretation of the inequality given in Theorem 10.3 is that if the redundancy increases the key equivocation will on average decrease and hence also the uncertainty about the key. In other words, redundancy reduction methods apparently increase the security of a cryptosystem. Another interpretation is that as long as $H(K) > D_L$ the key equivocation $H(K/C^L)$ cannot become equal to zero, so that on average it is not possible to determine the key unambiguously.

The same occurs if the length L of the cryptogram is small. This is highlighted by assuming the message source to be memoryless, which implies that

$$H(M^L) = LH(M). \qquad (10.17)$$

That is, the amount of information per message is equal to L times the amount of information per symbol.

If we substitute equation (10.17) into formula (10.13) this then leads to

$$H(K/C^L) \geq H(K) + L\ [H(M) - \log(\varepsilon)].$$ (10.18)

Thus the key equivocation can on average only become zero if

$$L \geq H(K)/\{\log(\varepsilon) - H(M)\}.$$ (10.19)

In other words, if the amount of information of the source, i.e. the uncertainty, is small only a few symbols are necessary to find the key. In practice, the user should therefore make certain that the information of the messages is as large as possible, by applying source coding for example. The value of L for which equality holds in formula (10.19) is called the *unicity distance*, denoted by UD. It is the minimum length of the ciphertext required to be able to find the key.

It should be remarked that the unicity distance indicates that there is a value for L whereby $H(K/C^L)$ can be equal to 0. This does not mean that the key can then be found. Here, as in the rest of this chapter, the approach has been based on the average case. This is inherent in the use of concepts within information theory, which concerns average amounts of information.

10.6 Exercises

10.1. Consider a general substitution cipher system with equally probable keys, where messages M^L with length L, designated as plaintext, are transformed, with the aid of a key K, into ciphertexts C.

Assume the information source is memoryless, and generates symbols from an alphabet $U = (u_1, u_2, \ldots, u_8)$ with probability of occurring

$$P = (\frac{1}{2}, \frac{1}{4}, \frac{1}{8}, \frac{1}{16}, \frac{1}{32}, \frac{1}{64}, \frac{1}{128}, \frac{1}{128}).$$

a. Calculate the amount of information in the plaintext. Calculate also the amount of information in the ciphertext.
b. Calculate the amount of information in the key.
c. Calculate the unicity distance.
d. Calculate the unicity distance if a Caesar substitution is used instead of a general substitution.

10.2. Consider a general cipher system. At the transmitter a plaintext M is transformed into a ciphertext C by a transformation T, which depends on a key K.

Assume that a memoryless information source is used which generates plaintexts consisting of symbols from the alphabet $V = (v_1, v_2, v_3, v_4)$. The probabilities of the symbols are given by

$$p_1 = \frac{7}{16}, p_2 = \frac{3}{16}, p_3 = \frac{2}{16}, p_4 = \frac{4}{16}.$$

For the encipherment a monoalphabetic substitution is used. It is assumed that each substitution alphabet has the same probability of occurring.

a. Calculate the amount of absolute redundancy in the plaintext and the amount of information in the key.

b. Calculate the unicity distance for this case and give an interpretation of the result.

Assume now that for the encipherment of the plaintext two consecutive symbols of the plaintext are considered as one new symbol. The substitution alphabet becomes $U = (u_1, u_2, u_3, \dots)$.

c. Assess the size this substitution alphabet should have. Calculate both the amount of information in the key and the redundancy of the plaintext.

d. Calculate the unicity distance in this case and explain the difference with the unicity distance as calculated in b.

10.3. Assume a source generates plaintexts consisting of letters from the normal alphabet consisting of 26 letters. It is assumed that the amount of information in the plaintext is the same as the entropy of the English language (assume 1.5 bits/symbol). To generate the ciphertext a poly-alphabetic substitution with a Vigenère-tableau is used, with a key length of 7 letters.

a. Calculate the amount of information in the key from the cryptanalyst's point of view, in the case where he knows that the key word consists of 7 letters, but he does not know which.

b. The same as a but now it is assumed that the 7 letters do not have to differ.

c. Calculate the unicity distance in the case of a.

d. Calculate also the unicity distance in the case of b. Explain the difference.

10.7 Solutions

10.1. *a.* Since the source is memoryless and thus $H(M^L) = LH(M) = LH(U)$ it suffices to consider the amount of information at the symbol level.
For the plaintext we obtain

$$H(M) = H(U) = -\sum_{i=1}^{8} p(u_i) \log p(u_i) \text{ bits/symbol.}$$

$$= -\frac{1}{2}\log\frac{1}{2} - \frac{1}{4}\log\frac{1}{4} - \frac{1}{8}\log\frac{1}{8} - \frac{1}{16}\log\frac{1}{16} - \frac{1}{32}\log\frac{1}{32}$$
$$- \frac{1}{64}\log\frac{1}{64} - 2\cdot\frac{1}{128}\log\frac{1}{128} = 1.98 \text{ bits/symbol.}$$

Since for the substitution cipher a one-to-one correspondence exists between the symbols of the plaintext and those of the ciphertext it is the case for the amount of information in the ciphertext that

$$H(C) = H(M) = 1.98 \text{ bits/symbol.}$$

b. The number of keys is 8! (or more precisely 8! – 1, excluding the key that leads to a ciphertext identical to the plaintext).
The information of the key yields

$$H(K) = \log 8! = 15.29 \text{ bits.}$$

c. The unicity distance equals

$$UD = \frac{H(K)}{\log(\varepsilon) - H(M)}$$

where ε is the source alphabet size.
We find

$$UD = \frac{15.29}{\log 8 - 1.98} \approx 15.$$

It is the average minimum number of symbols of the ciphertext needed for finding the key from the ciphertext.

d. The number of keys for a Caesar substitution is 8. Now, the unicity distance decreases to

$$UD = \frac{\log 8}{\log 8 - 1.98} \approx 3.$$

10.2. *a.* For the information of the plaintext it is the case that

$$H(M) = H(V) = -\frac{7}{16}\log\frac{7}{16} - \frac{3}{16}\log\frac{3}{16} - \frac{2}{16}\log\frac{2}{16} - \frac{4}{16}\log\frac{4}{16}$$
$$\approx 1.85 \text{ bits/symbol.}$$

The absolute redundancy equals

$$D = \log(\varepsilon) - H(M) = \log 4 - 1.85 \approx 0.15 \text{ bit/symbol.}$$

The number of keys is 4! (or 4! – 1) and thus since they are equiprobable

$$H(K) = \log 4! = \log 24 \approx 4.58 \text{ bits.}$$

b. The unicity distance becomes

$$UD = \frac{H(K)}{\log(\varepsilon) - H(M)} = \frac{4.58}{0.15} \approx 31.$$

c. Since the original alphabet exists of four symbols, combining two symbols will lead to 16 new compound symbols.

The size of the substitution alphabet should be 16, now. And thus for the amount of information in the key we find

$$H(K) = \log 16! = 44.3 \text{ bits.}$$

In general, the absolute redundancy for messages of length L in case of memoryless sources is

$$D_L = L \log(\varepsilon) - H(M^L).$$

In the present case we find

$$D_2 = 2\{\log - H(M)\} = 2D = 0.30 \text{ bit/message.}$$

d. Combining pairs of symbols leads to a unicity distance

$$UD = \frac{H(K)}{D_L} = \frac{44.3}{0.30} \approx 148.$$

In order to compare this result that of b we should divide this unicity distance by 2, this yields 74, since now a symbol is twice as long as in the case of b. Clearly, the redundancy per symbol is not changed the number of keys has been increased.

10.3. *a.* Since the key consists of 7 different letters, the number of keys is $26 \cdot 25 \cdot 24 \cdot 23 \cdot 22 \cdot 21 \cdot 20 = 26!/21!$ and therefore, the amount of information in the key is

$$H(K) = \log \frac{26!}{21!} \approx 31.63 \text{ bits.}$$

b. In the general case we find

$$H(K) = \log 26^7 = 32.90 \text{ bits}$$

c. For the unicity distance we find

$$UD = \frac{H(k)}{\log(\varepsilon) - H(M)} = \frac{H(K)}{\log 26 - 1.5} = \frac{H(K)}{3.2}.$$

d. We obtain in the case of a $UD_a = 31.63/3.20 \approx 10$ and in the case of b $UD_b = 32.90/3.20 \approx 11$.

Since the key space of a is smaller than that of b, in general fewer symbols are necessary in order to find the key. Thus $UD_a \leq UD_b$.

Bibliography

Abramson, N. (1963), *Information theory and coding*, McGraw-Hill Book Company, New York.

Aczel, J. and Z. Daroczy (1975), *On measures of information and their characterization*, Academic Press, New York.

Ash, R.B. (1965), *Information theory*, Interscience, New York.

Azimoto, S. (1971), Information theoretical considerations on estimation problems, *Inform. Contr.* Vol. 19, pp. 181–194.

Bell, D.A. (1962), *Information theory and its engineering applications*, I. Pitman Ltd, London.

Berger, T. (1971), *Rate distortion theory: a mathematical basis for data compression*, Prentice-Hall, Englewood Cliffs, NJ.

Blahut, R.E. (1987), *Principles and practice of information theory*, Addison-Wesley, Reading, Mass.

Blahut, R.E. (1990), *Digital transmission of information*, Addison-Wesley, Reading, Mass.

Boekee, D.E. and J.C.A. van der Lubbe (1988), *Informatietheorie*, Delftse Uitgevers Maatschappij, Delft.

Chaundy, T.W. and McLeod, J.B. (1960), On a functional equation, *Proc. Edinburgh Math. Soc. Notes*, 43, pp. 7–8.

Cover, Th.M., and J.A. Thomas (1991), *Elements of information theory*, Wiley, New York.

Csiszár, I., Körner, J. (1981), *Information theory*, Academic Press, New York.

Daroczy, Z. (1970), Generalized information functions, *Inform. Contr.*, Vol. 16, pp. 36–51.

Fano, R.M. (1961), *Transmission of information: a statistical theory of communication*, Wiley, New York.

Feinstein, A. (1958), *Foundations of information theory*, McGraw-Hill, New York.

Feller, W. (1957), *An introduction to probability theory and its applications*, Wiley, New York.

Gallager, R.G. (1968), *Information theory and reliable communication*, Wiley, New York.

Goldman, S. (1953), *Information theory*, Prentice-Hall, Englewood Cliffs, NJ.

Guiasu, S. (1976), *Information theory with applications*, McGraw-Hill, New York.

Hartley, R.V.L. (1928), Transmission of information, *Bell Syst. Tech. J.*, Vol. 7, pp. 535–563.

Jelinek, F. (1968), *Probabilistic information theory*, McGraw-Hill, New York.

Lubbe, J.C.A. van der (1981), *A generalized probabilistic theory of the measurement of certainty and information* (PhD thesis), Delft University of Technology, Dept. E.E., Information Theory Group.

McEliece, R.J. (1977), *The theory of information and coding*, Addison-Wesley, Reading, Mass.

McMillan, B. (1953), The basic theorems of information theory, *Ann. Math. Statist.*, pp. 196–219.

Nyquist, H. (1924), Certainty factors affecting telegraph speed, *Bell Syst. Tech. J.*, Vol. 3, pp. 324–346.

Renyi, A. (1960), On measures of entropy and information, *Proc. Fourth Berkeley Symp. Math. Statist. and Prob.*, no. 1, pp. 547–561.

Shannon, C.E. (1948), The mathematical theory of communication, *Bell Syst. Tech. J.*, Vol. 27, pp. 379–423 and pp. 623–656.

Index